T0135963

Proceedings

of the

International Beilstein Symposium

on

Glyco-Bioinformatics

Bits 'n' Bytes of Sugars

October 4th – 8th, 2009

Potsdam, Germany

Edited by Martin G. Hicks and Carsten Kettner

BEILSTEIN-INSTITUT ZUR FÖRDERUNG DER CHEMISCHEN WISSENSCHAFTEN

Trakehner Str. 7 – 9
60487 Frankfurt
Germany

Telephone: +49 (0)69 7167 3211
Fax: +49 (0)69 7167 3219
E-Mail: info@beilstein-institut.de
Web-Page: www.beilstein-institut.de

IMPRESSUM

Glyco-Bioinformatics – *Bits 'n' Bytes of Sugars*, Martin G. Hicks and Carsten Kettner (Eds.), Proceedings of the Beilstein-Institut Symposium, October 4th – 8th 2009, Potsdam, Germany.

Bibliographic information published by the *Deutsche Bibliothek.*
The *Deutsche Bibliothek* lists this publication in the *Deutsche Nationalbibliografie*; detailed bibliographic data are available in the Internet at http://dnb.ddb.de.

ISBN 978-3-8325-2719-8

Layout by: Hübner Electronic Publishing GmbH
Steinheimer Straße 22a
65343 Eltville
www.huebner-ep.de

Printed by Logos Verlag Berlin GmbH
Comeniushof, Gubener Str. 47
10243 Berlin
www.logos-verlag.de

Cover Illustration by: Bosse und Meinhard
Kaiserstraße 34
53113 Bonn
www.bosse-meinhard.de

PREFACE

Glycomics is an emerging field within the -omics-sciences which addresses the investigation of the structure-function relationships of complex biosynthesized carbohydrates and the role they play within biological systems.

The post-genomic era has seen an explosion of activities in the areas of genomics and proteomics in both fundamental research as well as biotechnology applications. Sequencing and synthesis of nucleic acids and proteins has been performed in an automated fashion for many years now; these important basic techniques are now being applied to carbohydrates. Through the work of a number of key laboratories around the world, significant scientific and technical advances are being made resulting in an increasing number of data sets of important interactions of carbohydrates with proteins and nucleic acids become available.

The scientific progress of both genomics and proteomics relies on the interdisciplinary nature of a laboratory-based and a computer-based cooperation. Such an interdisciplinary "glyco"-based community is currently starting to become established; this symposium is aimed at supporting these efforts by bringing together glycochemists and biologists with experts in bioinformatics and computer sciences to lay the ground-work for a concerted effort in the area of glyco-bioinformatics.

The symposium covered the use of publicly available data, data mining, structure prediction and docking of carbohydrates, web-based services to combine proteomics and glycomics data for structure-function research and glycosylation analysis.

The meeting successfully paved the way towards the development of first ideas for the integration of glyco-bioinformatics in a universal platform that will serve biologists, chemists and all interested in glycosciences.

We would like to thank particularly the authors who provided us with written versions of the papers that they presented. Special thanks go to all those involved with the preparation and organization of the symposium, to the chairmen who piloted us successfully through the sessions and to the speakers and participants for their contribution in making this symposium a success.

Frankfurt/Main, December 2010

Martin G. Hicks
Carsten Kettner
Peter Seeberger

CONTENTS

Glyco-Bioinformatics – *Bits 'n' Bytes of Sugars*
October 4th – 8th, 2009, Potsdam, Germany

Glycoinformatics for Structural Glycomics

Stuart M. Haslam[1] **and David Goldberg**[2]

[1]Division of Molecular Biosciences, Imperial College London, London SW7 2AZ, U.K.

[2]Palo Alto Research Center, 3333 Coyote Hill Rd., Palo Alto, CA 94304, U.S.A.

E-Mail: [1]s.haslam@imperial.ac.uk, [2]goldberg@parc.com

Received: 26th February 2010 / Published: 10th December 2010

Abstract

Ultra-high sensitivity mass spectrometric strategies incorporating MALDI-MS/MS and nano-electrospray(ES)-MS/MS enable very complex mixtures of glycoproteins and glycolipids from biological extracts of cells and tissues to be screened thereby revealing the types of glycans present and, importantly, providing clues to structures that are likely to be functionally important. However, in contrast to the genomic and proteomic areas, the glycosciences lack accessible, curated and comprehensive data collections that summarize the structure, characteristics, biological origin and potential function of glycans that have been experimentally verified and reported in the literature. This lack of glycan databases has been identified by glycobiologists as the single biggest hindrance to their research. Additionally, the sparseness of glycan databases hampers the realization of bioinformatics tools for the interpretation of experimental data and the automatic determination of the glycan structure, therefore limiting the possibility of large scale glycomics studies. The current status of the field and possible future developments are outlined.

Introduction

Glycans, both in the form of polysaccharides or glycoconjugates (bound to proteins and lipids), are the most abundant class of biomolecules and are increasingly being implicated in human health. Glycosylation is by far the most important post-translational modification in terms of the number of proteins modified and the diversity generated. Since glycoproteins, glycolipids and glycan-binding proteins (GBPs, also called lectins, which specifically recognize particular glycan epitopes) are frequently located on the cell's primary interface with the external environment, the cell surface, many biologically significant events can be attributed to glycan recognition. For this reason the rapidly expanding glycoscience field is being increasingly recognized as an important component of life science research. In contrast to the genomic and proteomic areas, the glycosciences lack accessible, curated and comprehensive data collections that summarize the structure, characteristics, biological origin and potential function of glycans that have been experimentally verified and reported in the literature. This lack of glycan databases has been identified by glycobiologists as the single biggest hindrance to their research. Additionally, the sparseness of glycan databases hampers the realization of bioinformatics tools for the interpretation of experimental data and the automatic determination of the glycan structure, therefore limiting the possibility of large scale glycomics studies. The complexity of the glycan structures and the variety of techniques that are used for their study, pose additional obstacles to the development of a single automated tool that could have the same impact on glycomics as Mascot and SEQUEST have had for proteomics.

Mass Spectrometry, Glycomics and Glycoproteomics

Rapidly-increasing developments in the field of mass spectrometry, particularly over the past twenty years, have led to the achievements of new milestones regarding sensitivity and determination of molecular weight, with mass accuracies of 0.01% of the total molecular weight of the sample now routinely being attained. These features have extended the capabilities of mass spectrometers to the study of large biopolymers such as glycoproteins, which can be several hundreds of thousands of Daltons in mass. The ability to analyse minute quantities of sample within a complex mixture, together with enhanced sensitivity and accurate mass analysis has led to mass spectrometry becoming a method of choice for the analysis of carbohydrates. The objectives of a glycomics experiment are to define the complete complement of carbohydrate structures in a system. Depending on the experimental set up the system could be purified glycoproteins, SDS-PAGE gel bands, cells, tissues or organs biopsies or a complete organism such as *Caenorhabditis elegans*. The best strategy for such experiments is MALDI-TOF mass spectrometry analysis of permethylated derivatised glycan samples [reviewed in 1]. Such methodologies have been utilized by The Analytical Glycotechnology Core C of The Consortium for Functional Glycomics (CFG; http://www.functionalglycomics.org), which was established in 2001 with 'glue' grant funding from the National Institute of General Medical Sciences. The overarching goal of the

CFG is to define the paradigms by which protein–carbohydrate interactions mediate cell communication. The objectives of a glycoproteomics experiment are to define glycan populations at individual glycosylation sites in an individual glycoprotein. This is more complex and resource intensive, both in terms of equipment and man hours, than a glycomics experiment and is achieved by MALDI-TOF and/or ES-mass spectrometry analysis of glycopeptides. It can be greatly facilitated by prior glycomic analysis [reviewed in 2].

Glycoinformatics

The application of glycomic and glycoproteomic methodologies outlined above has lead to the generation of large volumes of carbohydrate structural data. The manual interpretation of such large data sets is time consuming and requires expert knowledge. This bottleneck in the process has causes a considerable slowing of progress. Compared to the field of proteomics the automated interpretation of MS spectra of glycans is still an evolving field. In the following sections the most powerful glycoinformatic tools for MS glycan data are described.

Cartoonist

Cartoonist is a family of programs that annotate mass spectra of glycopeptides or detached glycans. The annotation is done by labelling peaks of the spectrum with *Cartoons*, which are graphical representations of glycans. Cartoons are widely used in the glycobiology community, because they give a quick sense of a glycan's structure. They are especially useful when annotating a spectrum containing many different glycans – it is much easier to see trends and patterns in a page of graphical cartoons than in a page of chemical formulas.

The original version of Cartoonist was described in [3]. In this article we summarize some of the improvements and new applications since that original publication: specifically, automatic cartoon libraries, separate java program for displaying annotations, searching the CFG (Consortium for Functional Glycomics) database and principled setting of parameters. Although there are versions of Cartoonist for N-glycopeptides and O-glycans we will only discuss the version for detached N-glycans.

Automatic Cartoon Libraries

One component of Cartoonist is a large library of plausible cartoons. Since the publication of [3], we have automated library construction. The user only need specify a set of antennae. The default set of eleven is shown in Figure 1 on the left. The cartoon library is generated in two steps. In the first step, the antennae are placed into "slots" of a template N-glycan, to generate a large set of base cartoons (we also add a small number of cartoons to the base set which do not fit this pattern). In the second step, rules are applied to expand the base set to a much larger library of cartoons. Our current library-builder uses three rules and generates

145,010 cartoons from 15,256 base cartoons (with the default set of antennae). The rules are illustrated on the right hand side of Figure 1. The first rule takes a cartoon with a single fucose and generates a cartoon without the fucose. The second rule adds a bisecting GlcNAc to a cartoon. The third rule systematically substitutes one sialic acid for another. For example, if the base cartoon has two NeuAc monosaccharides, this rule will generate three additional cartoons: one with the first NeuAc substituted with a NeuGc, one with the second NeuAc substituted, and one with both NeuAcs substituted.

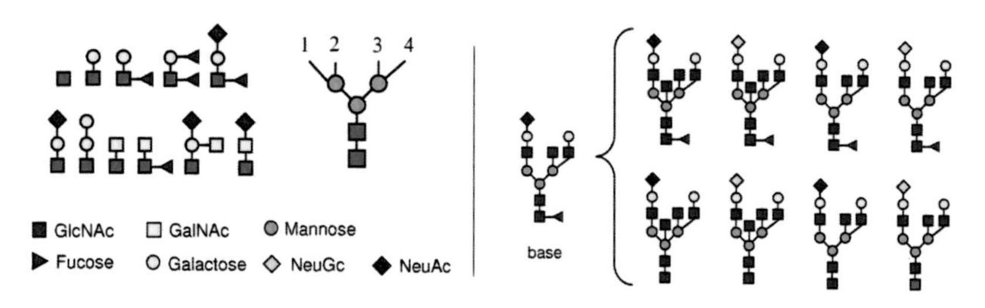

Figure 1. The standard Cartoon library is automatically generated using the 11 antennae shown on the left. Placing antennae into the slots numbered 1 – 4 gives the set of base cartoons. This set is further expanded using rules illustrated on the right.

Java Browser

The original version of Cartoonist was a monolithic program that computed annotations and produced a postscript file of the annotated spectrum which could be printed or displayed. But it was a static view of the spectrum. The current version has been modularized: the front end of Cartoonist produces a '.msa' file describing the annotations in human readable form. The back end is a Java browser, which reads the .msa file and displays it as an annotated spectrum. In addition to modularization, the advantage of this design is that the back end is a program that can pan and zoom through the spectrum and annotations. Figure 2 shows three different views of the browser on a spectrum of Human Monocytes from the CFG website. The first view shows the entire spectrum (Fig. 2A). There is not enough space to show all the annotations. The second view is zoomed in around Man-9 (2397 m/z), showing additional annotations (Fig. 2B). The third view illustrates one of the browser options: the cartoons can be magnified (or shrunk) (Fig. 2C).

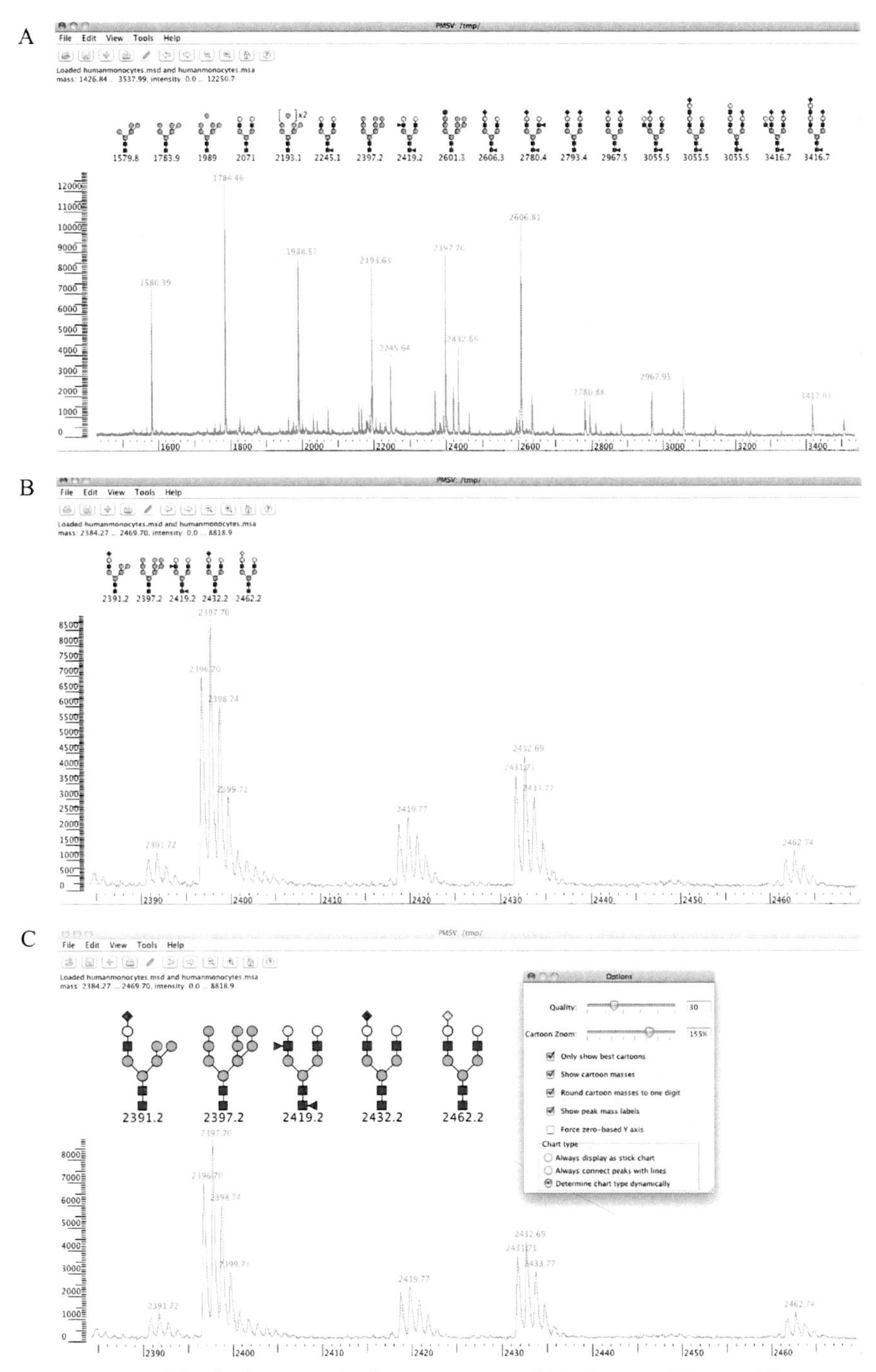

Figure 2. (A) The browser viewing a spectrum of N-glycans of Human Monocytes (from the CFG website). (B) The second view is zoomed in version of the first, the third view (C) has used the Options window to magnify the size of the cartoons.

Searching the CFG database

Cartoonist can be used not only to annotate spectra as they roll off of a mass spectrometer, but also to analyze existing databases of spectra. Motivated by the controversy about the existence of NeuGc in human cells and tissues [4] we used Cartoonist to search human samples in the CFG for NeuGc. Those spectra were annotated by human experts (without any NeuGc), but in this experiment we do not use that information, only the annotations made independently by Cartoonist. A systematic search found several matches, the most compelling of which were in Human Monocytes. Figure 3 shows the browser zoomed in on two possible NeuGc's in human monocytes. Each example has an additional peak with the correct m/z to be a NeuAc/NeuGc substitution, which is more evidence for the validity of the assignment to NeuGc.

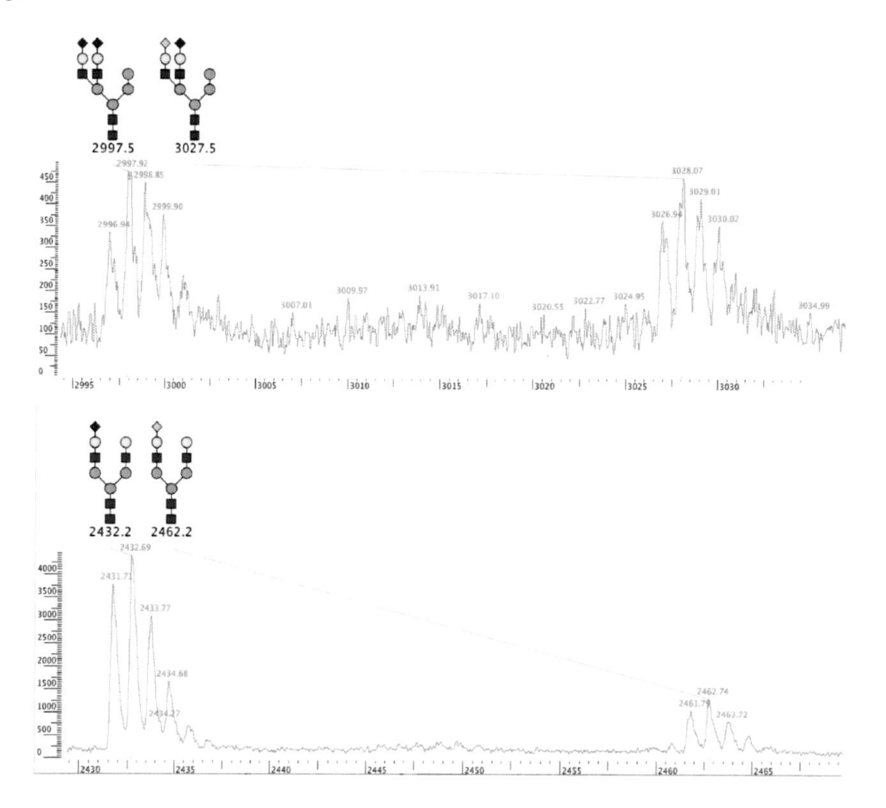

Figure 3. Peaks that may represent NeuGc in Human Monocytes. Principled Statistical Setting of Parameters

Cartoonist assigns a score to each annotation, which gives a rough sense of the probability that the annotation is correct. In the original version of Cartoonist, the score computation was rather *ad hoc*. The large set of human-expert annotations from the CFG web have been used to replace the original *ad hoc* setting of parameters in Cartoonist with statistically sound settings. One of the factors that make up the score is how close the observed isotope envelope is to the theoretical envelope. Cartoonist computes a number Δ that measures the difference in shape between the ideal and observed isotope envelopes. Figure 4 shows a histogram of Δ values for high-confidence annotations in the CFG spectra. By fitting this to a curve, we can convert Δ to a probability, specifically the "tail probability", that is, the area under the curve to the right of Δ. The blue dots lie on the best fitting curve of the form e^{-ax}, the red dots on xe^{-ax}. Clearly the latter is a better fit: specifically, the red curve is $(4/\mu^2)\ x\ e^{-(2x/\mu)}$, a probability distribution with mean of μ. For this data, $\mu = 0.18$. Using this curve gives an explicit formula for converting the Δ value of an isotope envelope to a probability.

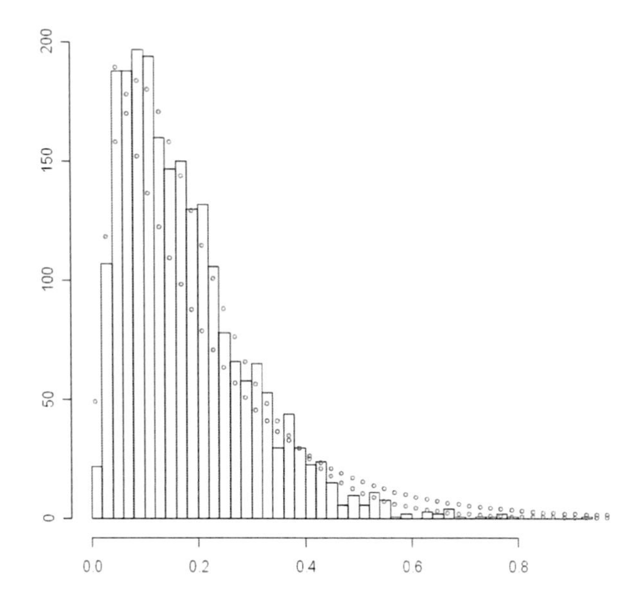

Figure 4. Histogram showing the distribution of Δ over a large set of isotope envelopes from the CFG spectra. The number Δ measures the discrepancy in shape between the observed and expected isotope envelopes. The red dots show the curve xe^{-ax} fitted to the histogram, the blue dots the curve e^{-ax}.

We also investigated whether a single Δ curve is sufficient, or whether we need to take into account how Δ varies with mass. To study the dependence on mass, we remove "outliers", the 10% of peaks with lowest intensity. For the remaining peaks the median mass is 2634. The mean of the Δ values is 0.17 both for peaks below the median and above the median. So it appears there is no gain to be had by having separate curves for different masses. However, as seen Figure 5 below, the high mass histogram is more ragged.

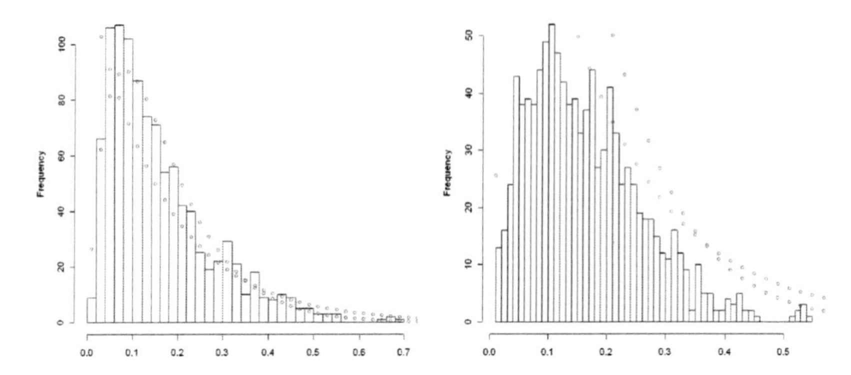

Figure 5. The histogram on left shows peaks with mass < 2600, on the right with mass ≥ 2600. For the left histogram, again the red curve is a better fit. Each histogram only considers peaks with rank < 557. In each case, $\mu = 0.18$. The red and blue curves are fit as in Figure 4.

Next we study deviation, which is the difference between the computed m/z and the observed (after recalibration) m/z. For judging the quality of a peak assignment, we would like to know the variance of the mass deviations, so that we can estimate how many "sigmas" the mass of an observed peak has strayed from the theoretical mass. That graph of this variance for the CFG spectra is shown in Figure 6.

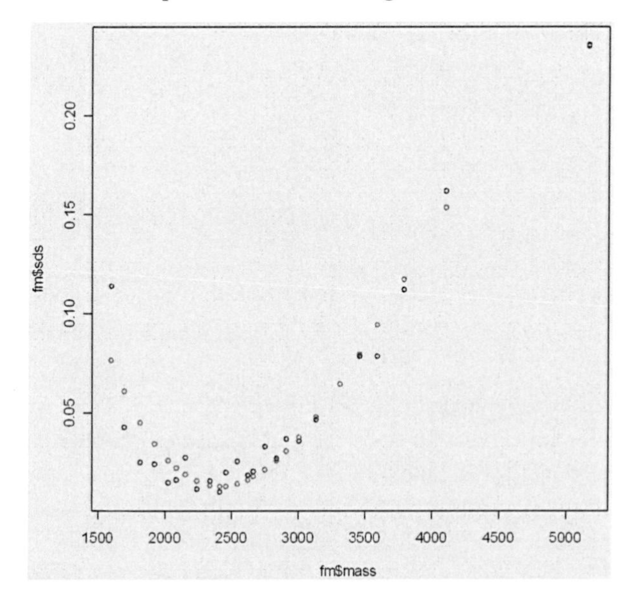

Figure 6. The histogram on left shows peaks with mass < 2600, on the right with mass ≥ 2600. For the left histogram, again the red curve is a better fit. Each histogram only considers peaks with rank < 557. In each case, $\mu = 0.18$. The red and blue curves are fit as in Figure 4.

The plot has grouped the peaks into 25 bins (of approximately 90 peaks each), and computed the standard deviation of the mass deviation for each bin. The black dots are the actual SD's, the red dots are the SD's fit to the cubic,

$$0.7443 - 0.0007138 \times mass + 0.0000002149 \times mass^2 - 0.00000000001855 \times mass^3$$

Cartoonist uses the data in Figure 4 and Figure 6 to compute a 'score' measuring the quality of an annotation. It makes the simplifying assumption that the deviation and isotope fit (Δ) are independent, and so their probabilities can be multiplied. The probability associated with Δ is the tail probability of the curve shown in Figure 4. The probability associated with the deviance is computed a bit differently. We assume the deviance has a Gaussian distribution, with a standard deviation σ given by Figure 6. So the probability associated with the deviation is again a tail probability, but this time the right hand tail of a Gaussian curve with standard deviation σ. The total probability is the product of these two numbers, and the final score is the minus logarithm of the probability.

EUROCarbDB

An Infrastructure Design Study (EUROCarbDB) was started in the European Union FP6 framework to lay out the foundations of a carbohydrate structural database and associated informatics tools. The partners of the EUROCarbDB design study have successfully established formats and standards for carbohydrate data exchange, developed tools to assist the interpretation of carbohydrate experimental data and designed a database architecture that can be used to store and retrieve glycan data. EUROCarbDB (http://www.ebi.ac.uk/eurocarb/) comprises an open-access relational database of glycan (carbohydrate) structures and primary research data, accessed via a web-based, browse/search/contribution interface. The database component is complemented by a suite of associated glyco-informatics tools, designed to aid in the elucidation and submission of glycan structures when used in conjunction with contemporary carbohydrate research workflows. Fundamentally, the database can be sub-divided into core and experimental components. The core database is comprised of four modules, which are interconnected. These four core modules are: sequence and structure, biological context, evidence, and reference, each of which is subsequently described. The relationship between each of these major modules is many-to-many; for example, a specific carbohydrate sequence may be found in (i. e. linked to) multiple species/ tissues, each of which may have several contributions of experimental evidence, such as MS, HPLC or NMR data or a single mass spectrum may be itself evidence for multiple carbohydrate sequences.

Two software tools, "Glyco-Peakfinder" and "GlycoWorkbench", have also been developed to assist in MS-carbohydrate structural data assignments and both freely provided by the EUROCarbDB initiative. Both tools can be used independently as they focus on different stages of the interpretation process of MS data. However, they can interact in a way that allows a complete and smooth workflow from raw data to a completely assigned spectrum and, furthermore, an easy to use way to collect all mandatory information for a new database entry. "Glyco-Peakfinder" is a web-application developed for *de novo* composition analysis of glycoconjugates. It is designed to ease the time intensive manual annotation of all kinds of MS spectra. Glycan profiles can be analysed as well as fragment spectra. "Glyco-Peakfinder" assigns all types of fragmentations including monosaccharide cross-ring cleavages (A-, B-, C-, X-, Y-, Z- fragments, gain and loss of small molecules). The tool provides full user control to handle modified glycans, including modifications at the reducing end, or of the whole structure. The option to calculate multiply-charged ions increases the range of application to mass spectrometric techniques other than just MALDI. Although the derived information for each entered m/z value is completely independent from the results of neighbouring peaks, a cross-linking of the results for several peaks provides additional sequence information. In addition the proposed glycan compositions can be used to search open access databases to assess if such a composition has previously been reported [5].

The further annotation process of MS/MS and MSn spectra can be performed by using the "GlycoWorkbench" suite of software tools. The graphical interface of "GlycoWorkbench" provides an environment in which structure models can be rapidly assembled, automatically matched with MS/MS and MSn data, and compared to assess the best candidate assignment. The main component of "GlycoWorkbench" is "GlycanBuilder", a flexible visual editor especially designed for a user-friendly input of glycan structures [6]. Glycans often exhibit tree-like non-linear structures, and their constituents exhibit great diversity. Because of the tree-like structure, the input of a glycan sequence is not as straightforward as writing a sequence of characters, as for DNA, RNA and peptide sequences. The lack of a suitable user-friendly graphical interface to input complex carbohydrate structures in a computer readable format has long been a severe deficiency in the practical application of glyco-related databases. Additionally, numerous alternative notations are commonly adopted to graphically represent glycan structures. After the desired input structures have been defined with "GlycanBuilder", the remaining components of "GlycoWorkbench" can be used to derive their fragments, compute the fragment masses, build a peak-list and annotate it. The computation of fragments and their masses from the intact structure is a central step for the annotation of MS/MS and MSn spectra. The "fragmentation tool" creates all topologically possible fragmentations of the precursor molecular ion, applying both multiple glycosidic cleavages and cross-ring fragmentations. The fragments are computed by recursively traversing the tree structure of the glycan and applying all the possible cleavages at each position. Fragmented structures are then subjected to the same process to produce multiple cleavages. For a given glycan fragment, the m/z ratio can be calculated both for native and derivatised structures (per-methylated/per-acetylated) taking into account several types and quantities of ion adducts. A visual editor of glycan fragments is also available, where the user can specify in which positions the cleavages are occurring on the displayed structure in order to reproduce an already known fragment molecule. The next step in the annotation process is the assignment of possible fragments to each m/z-value in a given peak list.

In "GlycoWorkbench" a peak list can either be loaded from a tab-separated text file, thus allowing for import from peak-picking software, or it can be created by typing mass and intensity values directly into the application. Once the peak-list is ready, the fragment m/z values from the *in silico* fragmentation are matched with a given accuracy to each peak in the list. The annotated peak-list can be displayed using various panels that show its different aspects. Each panel is based around a spreadsheet-like table view, whose cell values can be sorted by each column, and can be copied into spreadsheet applications. The annotated peak list can be stored in a specific XML format for later consultation or export for example into the EUROCarbDB database [7].

Acknowledgments

Thanks to Shane Ahern for his work in implementing the Java browser, Alessio Ceroni, Kai Maass and René Ranzinger for development of Glyco-Peakfinder and GlycoWorkbench. Funding: NIGMS (NIH Grant R01GM074128 to D.G.); the glycan analyses were performed by the Analytical Glycotechnology Core of the Consortium for Functional Glycomics (NIGMS GM62116 and the NCRR); the Biotechnology and Biological Sciences Research Council (BBSRC) Grant Nos. BBF0083091 and B19088 and the sixth European Union Research Framework Programme (EUROCarbDB RIDS Contract No. 011952).

References

[1] North, S.J., Hitchen, P.G., Haslam, S.M., Dell, A. (2009) Mass spectrometry in the analysis of N-linked and O-linked glycans. *Curr. Opin. Struct. Biol.* **19**:498 – 506. doi: http://dx.doi.org/10.1016/j.sbi.2009.05.005.

[2] Tissot, B., North, S.J., Ceroni, A., Pang, P.C., Panico, M., Rosati, F., Capone, A., Haslam, S.M., Dell, A., Morris, H.R. (2009) Glycoproteomics: past, present and future. *FEBS Lett.* **583**:1728 – 1735.
doi: http://dx.doi.org/10.1016/j.febslet.2009.03.049.

[3] Goldberg, D., Sutton-Smith, M., Paulson, J., Dell, A. (2005) Automatic annotation of matrix-assisted laser desorption/ionization N-glycan spectra. *Proteomics* **5**:865 – 875. doi: http://dx.doi.org/10.1002/pmic.200401071.

[4] Tangvoranuntakul, P., Gagneux, P., Diaz, S., Bardor, M., Varki, N., Varki, A., Muchmore, E. (2003) Human uptake and incorporation of an immunogenic nonhuman dietary sialic acid. *Proc. Natl. Acad. Sci. U.S.A.* **100**:12045 – 12050.
doi: http://dx.doi.org/10.1073/pnas.2131556100.

[5] Maass, K., Ranzinger, R., Geyer, H., von der Lieth, C.W., Geyer, R. (2007) "Glyco-peakfinder"- *de novo* composition analysis of glycoconjugates. *Proteomics* 7:4435 – 4444.
doi: http://dx.doi.org/10.1002/pmic.200700253.

[6] Ceroni, A., Dell, A., Haslam, S.M. (2007) The GlycanBuilder: a fast, intuitive and flexible software tool for building and displaying glycan structures. *Source Code Biol. Med.* 7:2 – 3.
doi: http://dx.doi.org/10.1186/1751-0473-2-3.

[7] Ceroni, A., Maass, K., Geyer, H., Geyer, R., Dell, A., Haslam, S.M. (2008) Glyco-Workbench: a tool for the computer-assisted annotation of mass spectra of glycans. *J. Proteome Res.* 7:1650 – 1659.
doi: http://dx.doi.org/10.1021/pr7008252.

Glyco-Bioinformatics – *Bits 'n' Bytes of Sugars*
October 4th – 8th, 2009, Potsdam, Germany

Mining Patterns from Glycan Structures

Ichigaku Takigawa[1,3], **Kosuke Hashimoto**[1,2], **Motoki Shiga**[1,3], **Minoru Kanehisa**[1] **and Hiroshi Mamitsuka**[1,3,*]

[1]Bioinformatics Center, Institute for Chemical Research, Kyoto University, Gokasho, Uji 611 – 0011, Japan

[2]DHHS/NIH/NLM, National Center for Biotechnology Information (NCBI), 8600 Rockville Pike, Bethesda, MD 20894, U.S.A

[3]Institute for Bioinformatics Research and Development (BIRD), Japan Science and Technology Agency (JST), Kawaguchi-shi, Saitama 332 – 0012, Japan

E-Mail: *mami@kuicr.kyoto-u.ac.jp

Received: 20th February 2010 / Published: 10th December 2010

Abstract

Glycans can be directed trees in which nodes are monosaccharides and edges are linkages extending between monosaccharides in the direction from the amino acid-connecting monosaccharide. We present an efficient method for mining frequent and statistically significant subtrees from glycan trees. The two key points of the method are: (1) It can reduce the number of redundant subtrees obtained by usual frequent subtree mining techniques and (2) can keep significant subtrees only by removing frequent but insignificant subtrees, like those with only one or a few nodes. We confirmed the efficiency of the approach in various manners, including biological significance. Our approach would be useful for mining unknown conserved patterns in larger glycan datasets to be obtained by a high-throughput manner in the future.

INTRODUCTION

Glycans or carbohydrate sugar chains are a major class of cellular macromolecules, working for a lot of important biological functions including antigen-antibody interactions [1] and cell fate controlling [2]. Like twenty types of amino acids for proteins and four types of bases for nucleic acids, building blocks of glycan structures can be monosaccharides, such as fructose, galactose, glucose and mannose [3]. Glycan structures are formed in that monosaccharides are connected to each other, allowing more than two branches extended from one monosaccharide but without any cycle. Glycans can be then *trees* in a computer science sense which consist of two parts: *nodes* and *edges* connecting nodes without forming any cycle. In addition, glycans can be *directed trees*, in which edges are directed, meaning that an edge is directed from a *parent* to a *child* [1]. This unique feature of glycans makes them different from other macromolecules such as nucleic acids and proteins: they are simple, non-branched, left-to-right type sequences of their building blocks. Furthermore this feature has made hard to determine the glycan structures experimentally and keep the size of structural databases on glycans relatively small and the speed of increasing the database size slow. However, due to long-term experimental efforts, the current number of structurally different glycans being accumulated in major glycan databases reaches about ten thousand, which might increase more by high-throughput techniques in the future [4]. This situation allows conducting data-driven techniques in bioinformatics for analyzing glycan structures and finding embedded biological significance [5, 6].

Our approach is based on techniques in data mining (or machine learning) to find rules, patterns or hypotheses from a large number of given examples. Machine learning techniques or problem settings can be classified into roughly three types: classification, clustering and pattern mining [7, 8]. Classification is to capture rules for distinguishing examples of each label (or class) from others, where labels are assigned to all examples, while clustering and pattern mining are conducted in each class. That is, clustering is to partition given examples into some groups and pattern mining is to find patterns common to given examples. Originally these techniques were applied to simple tables, where a row and a column correspond to an example and its attribute, respectively. A typical example in biology is a dataset of gene expression measured by cDNA microarray, where a row corresponds to some experimental condition while a column is a gene, each element in a table showing an expression value of the corresponding gene under the corresponding experimental condition [9]. Given a gene expression table, pattern mining means capturing rules which are genes, each being over- (or under-) expressed through given experimental conditions. The techniques to obtain this type of rules from a table have already been developed in a wide variety of manners and matured in some sense. The current focus in data mining and machine learning is to capture rules or patterns in more complex data, such as sequences, trees and graphs. Glycans can be directed trees, meaning that mining patterns from glycans can be along with the recent research topic of machine learning. The latest techniques for trees in machine learning are support vector machines with tree kernels [10] for classification, probabilistic models of trees [11 – 15] for clustering, and frequent subtree mining [16, 17] for pattern mining. Some conserved patterns are already known in glycans [18], and thus out of the three problem settings in machine learning, we focus on the third issue in this work, i. e. mining frequent subtrees from given trees.

In frequent subtree mining, a subtree is defined to be *frequent* if it appears more than a certain (pre-specified) number of times, which is called *minimum support* or *minsup*. For example, if we have ten trees and a subtree appears five times out of the ten trees, it is frequent if the minimum support is less than five. In this setting, there already exist efficient algorithms for finding frequent subtrees from a given set of trees, and any of these algorithms can be applied to glycans. However, a simple setting of frequent subtrees has two serious issues to which we address in this work. Firstly, resultant frequent subtrees are very redundant, because if a frequent subtree is found, all subtrees of the frequent subtree are frequent. Secondly, frequent subtrees are not necessarily found in a given set of trees significantly, because small subtrees, such as single nodes, are likely to be frequent, meaning that smaller subtrees are frequent in any dataset. In this work, we present an approach for dealing with these two problems.

Our approach has two parts, each responding to one of the two problems raised in the last paragraph (Note: originally our method appeared in [19]). In the first part, to reduce the redundancy of frequent subtrees, we developed a concept called α-closed frequent subtrees and an algorithm for mining α-closed frequent subtrees efficiently. In the second part, we prepare a dataset, generated randomly according to the distribution of a given set of trees, and employ hypothesis testing on both these two datasets to remove frequent but insignificant subtrees from the frequent subtrees obtained in the first part. With these two parts, we can relax the two serious issues in mining frequent subtrees.

We obtained frequent patterns of glycans by applying our approach to a real dataset of glycans. The frequent patterns generated include well-known, important patterns in glycobiology, the top pattern being peculiar to Type 2 oligosaccharides. We further evaluated our approach by the framework of classification, where we used patterns obtained by our approach as binary attributes of each glycan, an attribute value taking one if the corresponding pattern is in the glycan; otherwise zero. In this framework, a set of glycans can be a table of the attributes generated from the patterns, by which we can compare the discriminative performance of our approach with the most recent methods in machine learning, i. e. support vector machines with tree kernels. From this experiment, we showed that our approach could outperform competing methods by controlling the number of attributes or frequent patterns obtained.

METHOD

Our method has two parts, which can be clearly separated. However, we merged these two parts into one algorithm for efficiency. This point will be described in Part 3.

Part 1: Mining α-closed frequent patterns

The goal of the first part is to reduce the redundant outputs, i. e. frequent subtrees, and for this issue, we first need to introduce two important concepts: *closed* and *maximal* frequent subtrees [20]. Before that, we briefly define an important notion in frequent pattern mining: The number of appearances of subtree T in a given set of trees is called the *support* of T. Here, a frequent subtree T can be defined to be closed unless the support of any supertree of T is equal to that of T, while it can be maximal unless any supertree of T is frequent. More concretely, the idea of closed frequent subtrees allows to remove a redundant subtree if its support is equal to that of any of its supertrees. A problem of closed frequent subtrees is, however, that in general the support of a frequent tree is close to that of its supertree but not necessarily equal to. On the one hand, this means that in most cases closed frequent subtrees still suffer the redundancy problem of frequent patterns. On the other hand, the idea of maximal frequent subtrees outputs the largest frequent subtrees only, resulting in that many frequent subtrees can be summarized into only one maximal frequent subtree. Thus, maximal frequent subtrees can reduce the number of all frequent subtrees drastically, while a problem is that the support of each frequent subtree is lost. This means that maximal frequent subtrees might be too coarse to analyze all frequent subtrees. Thus, to solve these problems, we present an idea called α-closed frequent subtrees, which are a natural extension of closed frequent subtrees, relaxing the relatively strong restriction of closed frequent patterns to reduce the number of redundant patterns. The α-closed frequent subtrees can be defined in the following manner [19].

DEFINITION I
A frequent tree T is defined to be α-closed unless the support of any of its frequent supertrees is larger than or equal to $\alpha \times \text{support}(T)$, *where α takes a value between zero and one.*

We note that α-closed frequent subtrees have the following properties.

PROPERTY I
Given a set of trees and a minimum support, let $\mathcal{F}$ be the set of all frequent subtrees, $\mathcal{C}$ be the set of all closed frequent subtrees, $\mathcal{A}_\alpha$ be the set of all α-closed frequent subtrees, and $\mathcal{M}$ be the set of all maximal frequent subtrees. These sets satisfy that $\mathcal{M} \subseteq \mathcal{A}_\alpha \subseteq \mathcal{C} \subseteq \mathcal{F}$.

PROPERTY 2
$\mathcal{A}_\alpha$ is a monotone increasing family with respect to a, implying $\mathcal{A}_\alpha \subseteq \mathcal{A}_{\alpha'}$ $(\alpha \leq \alpha')$. If $\alpha = 1$, $\mathcal{A}_\alpha = \mathcal{C}$ and if $\alpha = 0$, $\mathcal{A}_\alpha = \mathcal{M}$.

PROPERTY 3
Once we are given a set of α-closed frequent subtrees, we can retrieve all frequent subtrees specified by the given α-closed frequent subtrees.

These properties show the nice features of α-closed frequent subtrees. PROPERTY 1 indicates the inclusive relationships among $\mathcal{M}$, $\mathcal{A}_\alpha$, $\mathcal{C}$ and $\mathcal{F}$, particularly the relationship among $\mathcal{A}_\alpha$, $\mathcal{C}$ and $\mathcal{F}$, being monotone, as shown in PROPERTY 2.

We further developed an efficient algorithm for enumerating all α-closed frequent subtrees from a given number of trees or glycans, keeping approximately the same computational burden of an algorithm for mining closed frequent subtrees. We do not go into the detail of the algorithm developed. Interested readers should refer [19], where we showed the idea behind the algorithm as well as the pseudo-code by which a software on mining α-closed frequent subtrees can be implemented.

Part 2: Ranking frequent subtrees by statistical testing

As mentioned in the Introduction, frequent patterns are not necessarily significant, since smaller patterns are likely to be frequent. Thus we checked the significance of each frequent pattern by using statistical testing over 'case' and 'control' datasets. The case dataset is a given set of trees, while the control dataset was generated by using the case dataset in the following two steps:

1. We compute the distribution of parent-child pairs in the case dataset.
2. We repeat the following procedure over all trees of the case dataset: For each tree, we replace each parent-child pair with another pair according the computed distribution, keeping the structure of the given tree, and save this new tree as a tree in the control dataset.

We then count, for each frequent subtree T, how many times T appears (and does not appear) in each of the case and control datasets, to make the following 2×2 contingency table.

Table 1. 2 × 2 contingency table.

	Case	Control	Total
T appears	N_{caT}	N_{conT}	N_T
T does not appear	N_{canT}	N_{connT}	N_{nT}
Total	N_{ca}	N_{con}	N

Finally, we can check the independence of the contingency table by one-sided Fisher's exact test. The probability that a contingency table is generated follows a hypergeometric distribution, which is given by using the counts in the contingency table as follows:

$$\mathrm{Pr} = \mathrm{N_T}!\ \mathrm{N_{nT}}!\ \mathrm{N_{ca}}!\ \mathrm{N_{con}}!/\mathrm{N_{caT}}!\ \mathrm{N_{canT}}!\ \mathrm{N_{conT}}!\ \mathrm{N_{connT}}!$$

The p-value of the one-sided Fisher's exact test on this table can be computed by the sum of all probabilities of tables that are more extreme than this table. We can rank frequent subtrees by the p-values of the corresponding contingency tables, according to their significance.

Part 3: Merging two parts for efficiency

In reality, the final output will be the top ranked patterns, say the top K patterns. For this case, we can take the following procedure: whenever a frequent pattern is obtained in the mining procedure, i.e. Part 1, we compute its p-value by using Part 2 and if the p-value is smaller than the largest one of the current lowest K p-values, we can keep this pattern; otherwise this pattern is discarded. By doing this, we can save the memory space as well as avoid the computational load caused by running the above two parts separately: generating all frequent patterns once and then sorting them by their p-values.

Experimental Results

Mining significant subtrees

We used all records in the KEGG Glycan database [21], removing entries of monosaccharides only and those containing nodes labelled by phosphorus (P) or sulphur (S). The total number of glycans we used for the case dataset was 7,454, and we generated a control dataset, which keeps the same size as that of the case dataset.

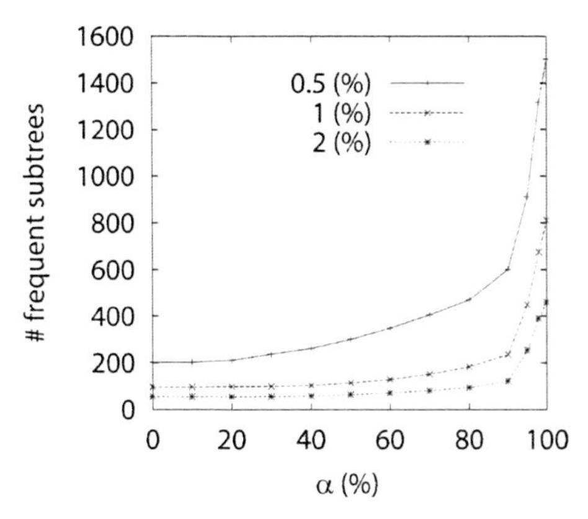

Figure 1. Mining significant subtrees in the KEGG Glycan Database: Frequent subtrees were obtained by our method, controlling the minimum support and the value of α.

Figure 1 shows the number of frequent subtrees obtained by our method, controlling the minimum support and the value of α. We note that here we did not use Part 2 or that the top *K* was set at a larger number than the number of frequent patterns obtained by the minsup in each setting. This figure indicates that the number of outputs was decreased by reducing the value of α (or increasing the *minsup*), meaning that redundancy of the output can be relaxed by the idea of α-closed frequent subtrees.

Rank		p-value	Support
1	Oβ1 ▲α1 >4,3 ■	1.6e-46	381
2	Oβ1—4■β1 Oβ1 >6,3 □	1.1e-40	164
3	Oβ1—3■β1—3Oβ1—4●	5.0e-26	109
4	Oβ1—3■β1—3O	5.6e-26	233
5	▲α1 Oβ1 >4,3 ■β1—3O	8.6e-26	83

▲ Fucose (Fuc)
○ Galactose (Gal)
□ N-acetylgalactosamine (GalNAc)
● Glucose (Glc)
■ N-acetylglucosamine (GlcNAc)

Figure 2. The top five significant subtrees, obtained by our method under the setting that *minsup* is 0.5% and α is 0.4.

Figure 2 shows the top five significant subtrees, obtained by our method under the setting that *minsup* is 0.5% and α is 0.4. We first note that they are not ranked by the number of appearances but by *p*-values. In fact, the third-ranked subtree has the support of 109, which is smaller than that of the fourth-ranked subtree, but has a lower *p*-value than that of the fourth. We found that all these five subtrees are biologically well-known, significant glycan motifs. For example, the first subtree is a common subtree among Lewis X, Lewis Y and sialyl Lewis X, which are all Type 2 oligosaccharides, being attached to the membrane of red blood cells and used for the categorization of human blood types. The second subtree is famous as the core of O-glycans, a major class of glycans, and similarly the third and fourth subtrees are typical core parts of another major class of glycans, called glycosphingolipids. These results indicate that our approach detected typical patterns or motifs in glycans, implying that relatively lower ranked subtrees might include unknown patterns in the literature.

Applying to classifying glycans in their classes

We then applied the obtained frequent patterns to the problem of classifying glycans to validate the significance of the obtained subtrees. In fact, once we obtain a set of frequent patterns, we can generate attributes of an arbitrary glycan by checking whether this glycan has each of the frequent patterns: If the glycan has a frequent pattern, the corresponding attribute takes one; otherwise zero. This means, if 1,000 frequent subtrees are obtained, we can assign 1,000 binary attributes to an arbitrarily given glycan.

The problem setting is that we regard 485 O-glycans in KEGG as positives (which used as the case dataset) and randomly synthesized glycans as negatives (which are used for the control dataset as well). We performed 10×10 cross-validation over these glycans, meaning that we repeated the following ten times: we divided the examples into ten blocks and used nine out of ten blocks for training and the rest for testing. The performance was measured by AUC (the area under the ROC curve) [22], a current standard measure for classification, and the final result was averaged over all 100 runs of the 10×10 cross-validation.

We chose three competing methods, all using support vector machines but different three tree kernels. The first one is convolution kernel [23], which enumerates all possible subtrees of given two trees and checks the number of subtrees shared between the two trees. The second one is co-rooted subtree kernel [10], which checks the similarity between given two trees in a similar manner to the convolution kernel but by using part of their all subtrees. The third one is 3-mer kernel [23], which focuses on subtrees with only three nodes for computing similarity between two trees.

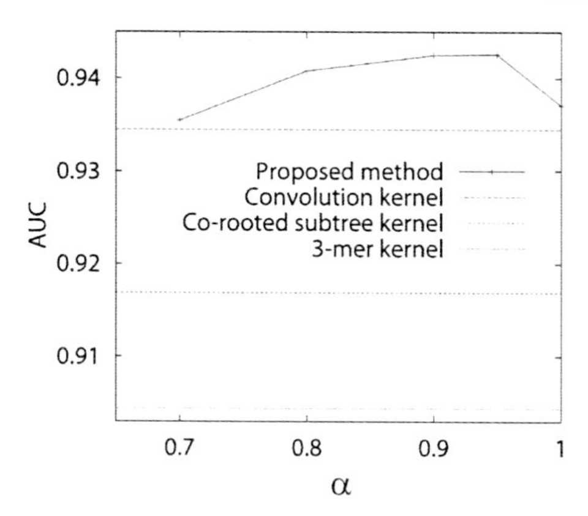

Figure 3. AUCs of the three competing methods and our proposed method for which *minsup* = 2.

Figure 3 shows the AUCs of the three competing methods and our proposed method for which *minsup* = 2. We note that the performance of our method can be controlled by changing α. For α of less than 0.7, the number of frequent patterns (or attributes) is too small to have a high AUC, while our method outperformed other three competing methods in AUC for α of no less than 0.7. In addition, we note that our method can show all attributes used in experiments as subtrees while other kernel-based methods cannot.

Concluding remarks

We have presented an approach for mining subtrees embedded in glycans which are frequent as well as statistically significant. We confirmed that the patterns ranked in the top five by our approach are consistent with the known literature in glycobiology. Experimental results further indicated the performance advantage of our method in classification. We did not check the scalability of our approach in our experiments, but the time- and space-scalability are also an advantage of our approach which would be more useful for larger-sized glycan databases to be obtained in the future. Our approach found the significant patterns removing redundancy. They are however still relatively similar to each other in some cases and needed to be summarized more in some form. Possible future work would be to develop a method for providing an overview of the significant patterns mined, showing their correlations with biological classification.

References

[1] Varki, A., Cummings, R., Esko, J., Freeze, H., Hart, G. and Marth, J. (1999) *Essentials of Glycobiology.* CSHL Press, New York (NY, USA).

[2] Stanley, P. (2007) Regulation of notch signaling by glycosylation. *Current Opinion in Structural Biology* **17**(5):530 – 535.
doi: http://dx.doi.org/10.1016/j.sbi.2007.09.007.

[3] Marth, J.D. (2008) A unified vision of the building blocks of life. *Nature Cell Biology* **10**(9):1015 – 1016.
doi: http://dx.doi.org/10.1038/ncb0908-1015.

[4] Haslam, S.J., North, S.J. and Dell, A. (2006) Mass spectrometric analysis of N- and O-glycosylation of tissues and cells. *Current Opinion in Structural Biology* **16**:584 – 591.
doi: http://dx.doi.org/10.1016/j.sbi.2006.08.006.

[5] Raman, R., Raguram, S., Venkataraman, G., Paulson, J.C. and Sasisekharan, R. (2005) Glycomics: an integrated systems approach to structure-function relationships of glycans. *Nature Methods* **2**(11):817 – 824.
doi: http://dx.doi.org/10.1038/nmeth807.

[6] Mamitsuka, H. (2008) Informatic innovations in glycobiology: relevance to drug discovery. *Drug Discovery Today* **13**(3 – 4):118 – 123.
doi: http://dx.doi.org/10.1016/j.drudis.2007.10.013.

[7] Bishop, C.M. (2006) *Pattern Recognition and Machine Learning*. Springer, New York (NY, USA).

[8] Han, J., Cheng, H., Xin, D. and Yan, X. (2007) Frequent pattern mining: current status and future directions. *Data Mining and Knowledge Discovery* **15**(1):55 – 86.
doi: http://dx.doi.org/10.1007/s10618-006-0059-1.

[9] Speed, T. (2003) *Statistical analysis of gene expression microarray data*. Chapman & Hall/CRC Press, Boca Raton (Florida, USA).
doi: http://dx.doi.org/10.1201/9780203011232.

[10] Shawe-Taylor, J. and Cristianini, N. (2004) *Kernel methods for pattern analysis*. Cambridge University Press, Cambridge (UK).

[11] Aoki, K.F., Ueda, N., Yamaguchi, A., Kanehisa, M., Akutsu, T. and Mamitsuka, H. (2004) Application of a new probabilistic model for recognizing complex patterns in glycans. *Bioinformatics* **20**(Supplement 1):i6-i14.
doi: http://dx.doi.org/10.1093/bioinformatics/bth916.

[12] Ueda, N., Aoki-Kinoshita, K.F., Yamaguchi, A., Akutsu, T., Mamitsuka, H. (2005) A probabilistic model for mining labeled ordered trees: capturing patterns in carbohydrate sugar chains. *IEEE Transactions on Knowledge and Data Engineering* **17**(8):1051 – 1064.
doi: http://dx.doi.org/10.1109/TKDE.2005.117.

[13] Aoki-Kinoshita, K.F., Ueda, N., Mamitsuka, H. and Kanehisa, M. (2006) Profile-PSTMM: Capturing tree-structure motifs in carbohydrate sugar chains. *Bioinformatics* **22**(14): e25-e34.
doi: http://dx.doi.org/10.1093/bioinformatics/btl244.

[14] Hashimoto, K., Aoki-Kinoshita, K.F., Ueda, N., Kanehisa, M. and Mamitsuka, H. (2006) A new efficient probabilistic model for mining labeled ordered trees. In: *Proceedings of the Twelfth ACM SIGKDD International Conference on Knowledge Discovery and Data Mining (KDD 2006)*, (Tina Eliassi-Rad, Lyle H. Ungar, Mark Craven and Dimitrios Gunopulos, eds.). pp. 177 – 186, ACM Press.

[15] Hashimoto, K., Aoki-Kinoshita, K.F., Ueda, N., Kanehisa, M. and Mamitsuka, H. (2008) A new efficient probabilistic model for mining labeled ordered trees applied to glycobiology. *ACM Transactions on Knowledge Discovery from Data* **2**(1):Article 6.

[16] Asai, T., Abe, K., Kawasoe, S., Arimura, H., Sakamoto, S. and Arikawa, S. (2002) Efficient substructure discovery from large semi-structured data. In: *Proceedings of the 2002 SIAM International Conference on Data Mining (SDM 2002)*, (Robert Grossman, Jiawei Han, Vipin Kumar, Heikki Mannila and Rajeev Motwani, eds.). pp. 158 – 174, SIAM.

[17] Zaki, M.J. (2002) Efficiently mining frequent trees in a forest. In: *Proceedings of the Eighth ACM SIGKDD International Conference on Knowledge Discovery and Data Mining (KDD 2002)*, (David Hand, Daniel Keim and Raymond Ng, eds.). pp. 71 – 80, ACM Press.

[18] Lanctot, P.M., Gage, F.H. and Varki, A.J. (2007) The glycans of stem cells. *Current Opinion in Chemical Biology* **11**:373 – 380.
doi: http://dx.doi.org/10.1016/j.cbpa.2007.05.032.

[19] Hashimoto, K., Takigawa, I., Shiga, M., Kanehisa, M. and Mamitsuka, H. (2008) Mining significant tree patterns in carbohydrate sugar chains. *Bioinformatics* **24**(16):i167-i173.
doi: http://dx.doi.org/10.1093/bioinformatics/btn293.

[20] Chi, Y., Xia, Y., Yang, Y. and Muntz, R.R. (2005) Mining closed and maximal frequent subtrees from databases of labeled rooted trees. *IEEE Transactions on Knowledge and Data Engineering* **17**(2):190 – 202.
doi: http://dx.doi.org/10.1109/TKDE.2005.30.

[21] Hashimoto, K., Goto, S., Kawano, S., Aoki-Kinoshita, K.F., Ueda, N., Hamajima, M., Kawasaki, T. and Kanehisa, M. (2006) KEGG as a glycome informatics resource. *Glycobiology* **16**: 63R-70R.
doi: http://dx.doi.org/10.1093/glycob/cwj010.

[22] Mamitsuka, H. (2006) Selecting features in microarray classification using ROC curves. *Pattern Recognition* **39**(12):2393 – 2404.
doi: http://dx.doi.org/10.1016/j.patcog.2006.07.010.

[23] Yamanishi, Y., Bach, F. and Vert, J.-P. (2007) Glycan classification with tree kernels. *Bioinformatics* **23**(10):1211 – 1216.
doi: http://dx.doi.org/10.1093/bioinformatics/btm090.

Bioinformatics – Key to the Future of Chemical Glycomics

Peter H. Seeberger

Max Planck Institute for Colloids and Surfaces, Potsdam, Germany,
Free University of Berlin, Arnimallee 22, 14195 Berlin, Germany, and
The Burnham Institute, La Jolla, CA, U.S.A.

E-Mail: peter.seeberger@mpikg.mpg.de

Received: 28th May 2010 / Published: 10th December 2010

Abstract

The glycome is more complex than either the genome or the proteome. Efforts to understand glycomics are producing information regarding the structure and function of carbohydrates. Branching and stereochemistry of the glycosidic linkage renders carbohydrates much more complex than oligonucleotides and proteins. Bioinformatics is a key technology to extract the information relayed via glycans. Three major classes of mammalian carbohydrates, glycolipids, O- and N-linked glycans, were analyzed based on the largest available database. The average oligosaccharide is composed of about eight monosaccharide units and while about a quarter of all oligosaccharides are strictly linear, the remainder are branched at least once. Glucosamine, galactose and mannose are dominating and comprise about 75% of the monosaccharides within mammalian oligosaccharide frameworks. α-linked sialic acid, α-linked fucose and β-linked galactose decorate the majority of reducing termini. Glucose as the most abundant carbohydrate in mammals plays only a very minor role within these structures. Particular emphasis was placed on analyzing the way the monosaccharide units are linked within the oligomeric framework. Just eleven monosaccharide connections account for more than 75% of all linkages. Thus, the number of structural combinations found in nature – the part of the occupied mammalian glycospace – is much smaller than expected. Only 36 monosaccharide building blocks are required to construct 75% of the 3299 mammalian oligosaccharides.

INTRODUCTION

Three major repeating biomacromolecules, polynucleotides, polypeptides and carbohydrates, are responsible for much of the information transfer in biological systems. Encoding and transmission of information relies on the construction of diverse macromolecules that contain the message. Polynucleotides serve as the blueprint of life in form of DNA; polypeptides carry out most of reactions in living cells. Both polymers are strictly linear and derived biosynthetically via reliable templated syntheses. DNA is composed of four nucleotides and mammalian proteins have 20 proteinogenic amino acids that determine polymer diversity: 4096 (4^6) hexanucleotides and 64 million (20^6) hexapeptides are possible. Posttranslational modifications such as phosphorylation, glycosylation, and lipidation further increase protein complexity.

The term "carbohydrates" describes a host of different bio-oligomers composed of monosaccharides. Oligosaccharides are almost always part of glycoconjugates, i.e. the combination of a sugar chain with a protein (glycoprotein), a lipid (glycolipid) or both lipid, and protein (glycosylphosphatidylinositol (GPI) anchored proteins) [1]. Carbohydrate chains can be branched, since each monosaccharide provides different positions around the ring that can be connected. In contrast to amide or phosphate diester linkages, the formation of each glycosidic linkage creates one new stereogenic centre. Carbohydrate complexity is increased by the stereocentres that constitute the ring in addition to ring size, linkage position, branching as well as further attachments such as sulfation, methylation, and phosphorylation.

Not surprisingly, carbohydrate complexity dwarves that of both DNA and proteins but to date has been assessed on a purely theoretical level [2]. We performed calculations regarding the diversity of mammalian carbohydrate structures, based only on the "ten mammalian monosaccharides" (Glc, Gal, Man, Sia, GlcNAc, GalNAc, Fuc, Xyl, GlcA, IdoA) not considering any further attachments. The number of structural combinations encountered in nature – the part of the glycospace that is actually occupied – has not yet been elucidated. A systematic structural analysis of mammalian oligosaccharide structures deposited in glycan databases will aid our understanding of carbohydrate diversity and help to identify a putative set of monosaccharide building blocks for efficient carbohydrate assembly.

Table 1. Diversity space of oligonucleotides, peptides and mammalian oligosaccharides. The numbers for the mammalian oligosaccharides are based on the "ten mammalian monosaccharides": D-Glc [4], D-Gal [4], D-Man [4], D-Sia [4], D-GlcNAc [3], D-GalNAc [3], L-Fuc [3], D-Xyl [3], D-GlcA [3], L-IdoA [3]. The number of substitutable OH groups (excluding the anomeric one) is given in square brackets. Commonly, only the pyranose ring forms, but not the furanose ring forms of the above mentioned monosaccharides are found in mammals [20].

Oligomer size	Numbers of different oligomers		
	Nucleotides	Peptides	Carbohydrates
1	4	20	20
2	16	400	1 360
3	64	8 000	126 080
4	256	160 000	13 495 040
5	1 024	3 200 000	1 569 745 920
6	4 096	64 000 000	192 780 943 360

Access to pure, structurally defined carbohydrates remains difficult at a time when the automated synthesis of oligonucleotides [3] and oligopeptides [4] is common. While biologically relevant oligosaccharides can be assembled from monosaccharides in a linear fashion on an automated synthesizer [5, 6], no general method for non-specialists to draw from a set of commercially available building blocks exists yet. The structural complexity of carbohydrates may complicate a comprehensive synthesis approach in case too many building blocks are needed.

A better understanding of the structures actually found in nature can guide the selection of building blocks needed for assembly. A database of reliable structures is required to analyze carbohydrate diversity. Relatively limited data sets exist, since carbohydrate isolation and structure elucidation are formidable challenges. The systematic collection of carbohydrates in databases is lagging far behind genomics and proteomics. Currently no database provides access to all published glycan structures although several commercial and publicly funded initiatives are working to make glycan structures available in a well-structured and annotated digital representation [7]. These databases contain mainly information about O- and N-linked glycans since their isolation and sequencing is more tractable than that of glycosaminoglycans for example. Reported here is a detailed statistical analysis of the GLYCOSCIENCES.de database [8] to elucidate the mammalian glycospace with a focus on the oligosaccharide portions of O- and N-glycans.

Glycan Database Analysis

The complexity analysis was based on 3299 oligosaccharides from 38 mammals (9). Non-carbohydrate portions such as amino or fatty acids at the reducing terminus were not considered. Most oligosaccharides (2128 of 3299, 64%) are of human origin, the rest are derived from cow, rat, pig, mouse and other species. Since the statistical analyses did not reveal any relevant difference between the human and the mammalian set of oligosaccharide structures we focused our analysis on the mammalian sugars.

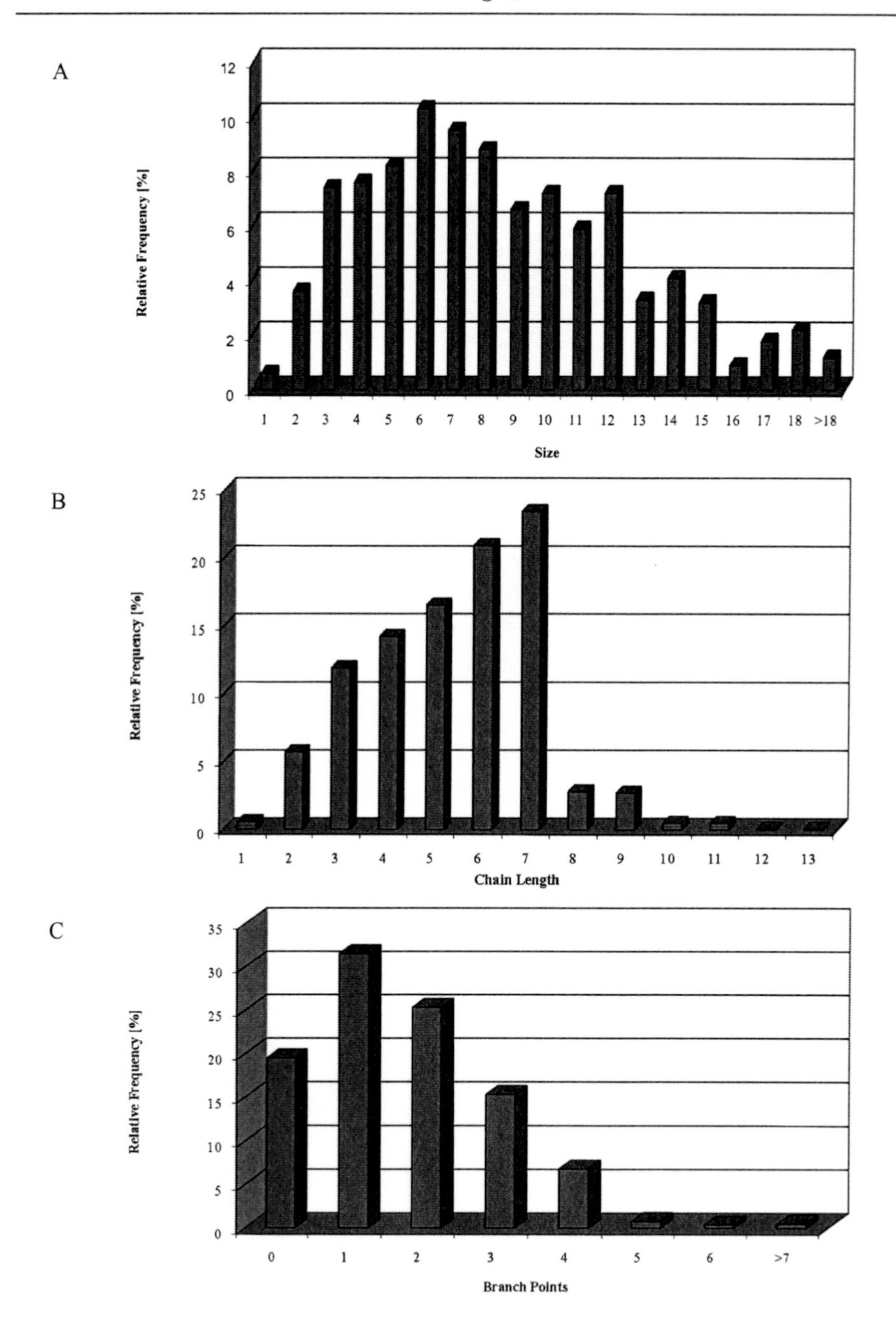

Figure 1. (A) Size of mammalian O- and N-glycans. The relative frequency is given in %. The largest oligosaccharide contains 37 monosaccharides; (B) Chain length of mammalian O- and N-glycans. The longest chain contains 13 monosaccharide moieties; (C) Branching complexity of mammalian O- and N-glycans.

Initially we addressed some basic questions: What is the most common size of an oligosaccharide? What is the typical chain length within a branched oligosaccharide? What portion of oligosaccharides is linear, what portion is branched? The size of an oligosaccharide is described as the number of monosaccharide units that make up the oligomer. The chain length is the longest path of monosaccharide units from the reducing end to the non-reducing terminus of the chain. The number of terminal residues was calculated as well. It differs by one from the number of branch points.

The average oligosaccharide is composed of about eight monosaccharide units whereas the sugars in the database vary in size from one (*e.g.* T_N antigen) to 37 monosaccharide units (Figure 1a). Most mammalian structures (about 95%) have a shorter chain length than eight residues (Figure 1b). The longest mammalian carbohydrate structure in the database has a maximum chain length of 13. About one fifth of the oligosaccharide structures in the database are linear. More than half of all oligosaccharides are branched once or twice, while 22% of the structures are branched three or four times. Few carbohydrates are branched five times or more than five times with a maximum of nine branch points (Figure 1c).

N-Acetyl-D-glucosamine (32%), D-galactose (25%) and D-mannose (19%) comprise more than 75% of all monosaccharide units found in mammalian oligosaccharides. Sialic acid and L-fucose are found less frequently (8% each). The most abundant monosaccharide in nature, D-glucose, that makes up cellulose, starch and glycogen, astonishingly plays only a minor role within mammalian O- and N-glycans. Three different monosaccharides dominate the non-reducing terminus, the site most often recognized by carbohydrate-binding proteins: D-galactose, sialic acid and L-fucose each cap about a quarter of oligosaccharides. D-Mannose and *N*-acetyl-D-glucosamine each terminate about 8% of mammalian oligosaccharides.

In planning a comprehensive, general and linear synthetic approach a set of building blocks containing the proper protective groups to install all possible connectivities and stereogenic centres is mandatory. Different protective groups mark hydroxyl groups that serve as nucleophiles during chain extension and those that remain latent. The protective groups used should also control the stereochemical outcome of glycosylation reactions. To further complicate matters, the sterics, electronics and conformation of the monosaccharides are fundamentally influenced by the choice of protective groups [10]. One aim of this study is to derive a minimal set of putative monosaccharide building blocks required to assemble the majority of mammalian oligosaccharides in a strictly linear fashion. Procurement of monosaccharide building blocks is a formidable challenge [11] and a defined number of reliable standard components for oligomer construction would help in synthetic planning and for practical reasons. These building blocks would be utilized in the build-up of linear and branched molecules by solution- and solid-phase methods.

Each of the ten mammalian monosaccharide units can in principle be connected to its neighbours in a variety of different ways including different anomeric configurations as well as branching once or more often. To construct all theoretically possible mammalian oligosaccharides by linear chemical synthesis 224 different building blocks would be required (for further details see SI). Due to this large number, special care was taken to elucidate the stereochemistry at the anomeric position (α or β) and the position of linkages within mammalian O- and N-glycans. The results illustrated in Figure 2 are stunning: 80% of the monosaccharide linkages within the oligomers can be constructed using only 13 building blocks. The most frequently occurring connections are 4-linked β-GlcNAc, capping α-Sia, capping α-Fuc, capping β-Gal, 2-linked α-Man and 3-linked β-Gal.

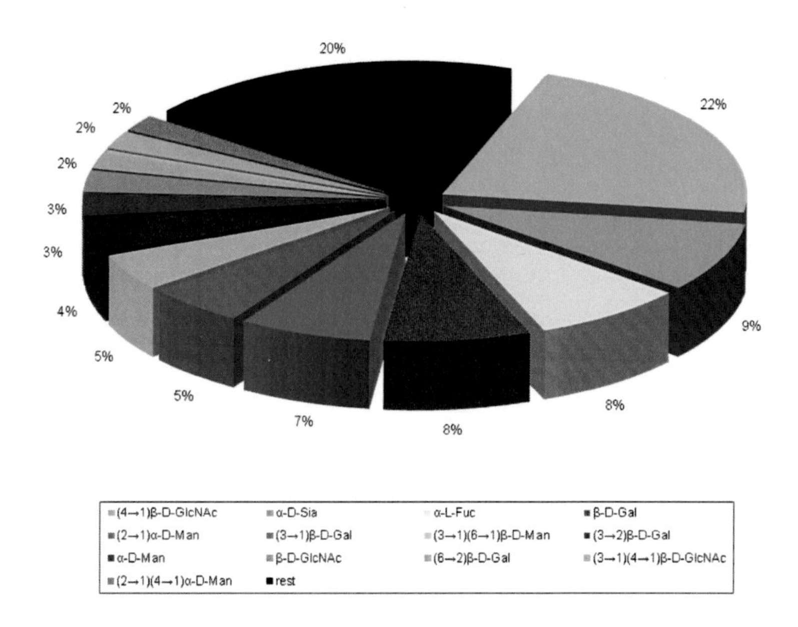

Figure 2. The 13 most abundant monosaccharide units (with linkage mode and position) found in mammalian O- and N-glycans.

Based on this analysis, key building blocks needed for the construction of mammalian oligosaccharides can be designed. Twenty putative building blocks **1**–**20** have been postulated to obtain the most abundant linkages. For this purpose, benzyl groups (Bn) were selected for the permanent protection of hydroxyls. Where participating groups are needed to ensure anomeric specificity, pivaloyl (Piv) [12], acetyl (Ac) or benzoyl (Bz) groups are placed for permanent protection. Fluorenylmethoxycarbonyl (Fmoc) was selected as temporary protecting group and also serves as a participating group in the C2 position for temporary protection [13]. To install branched carbohydrate structures two or more temporary protecting groups that can be cleaved chemoselectively are necessary. Levulinoyl ester

(Lev) and *p*-methoxybenzyl (PMB) were selected as other temporary protecting groups [14]. Similar protecting group schemes can be selected alternatively. However, the large majority of the building blocks presented in Figure 3 have been tested successfully for their utility in solution- and solid-phase oligosaccharide syntheses [15]. The sialic acid and β-mannose building blocks **2** and **7** represent a special challenge concerning selective glycosidic bond construction. So far, α-sialic acid and β-mannosidic bonds were constructed using disaccharide modules in place of monosaccharide units [16]. These challenges are currently being addressed.

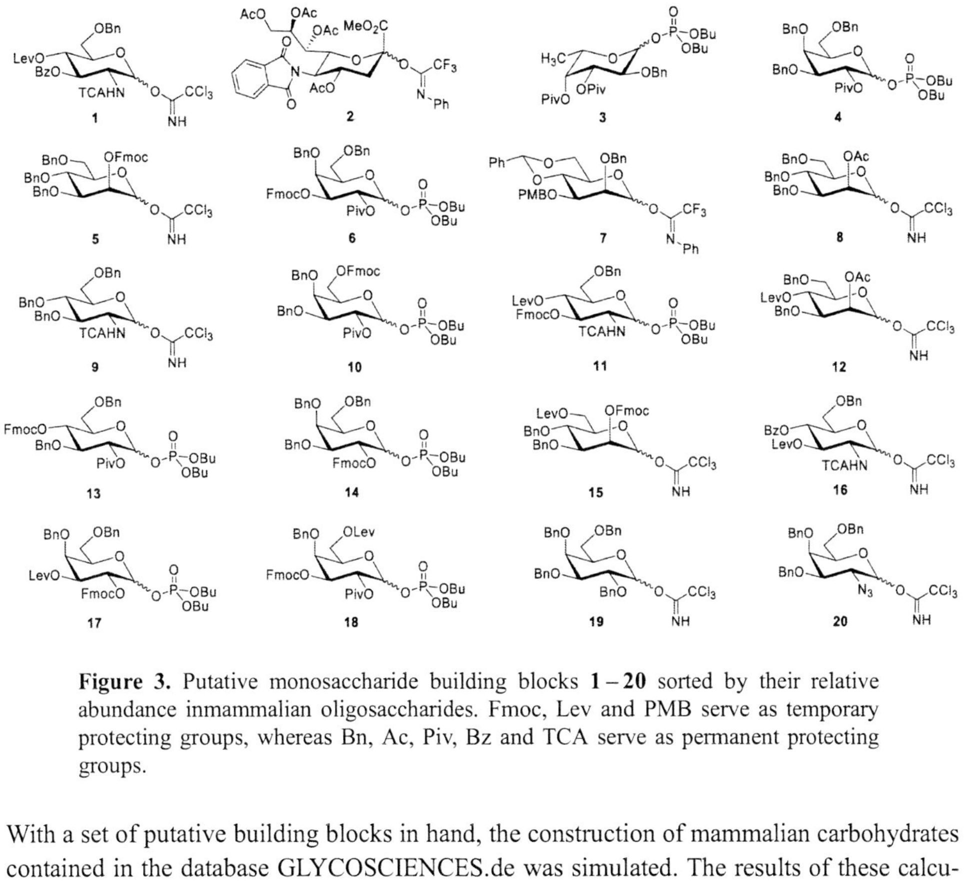

Figure 3. Putative monosaccharide building blocks **1 – 20** sorted by their relative abundance inmammalian oligosaccharides. Fmoc, Lev and PMB serve as temporary protecting groups, whereas Bn, Ac, Piv, Bz and TCA serve as permanent protecting groups.

With a set of putative building blocks in hand, the construction of mammalian carbohydrates contained in the database GLYCOSCIENCES.de was simulated. The results of these calculations are impressive: about 60% of the 3299 mammalian oligosaccharides are accessible with only 25 building blocks (Figure 4)! Just eleven more building blocks are needed to construct 75% of oligosaccharides. To produce 90% of all structures, a set of 65 building blocks is required. The number of building blocks to access the last 10% of mammalian oligosaccharides increases tremendously. The occurrence of rare monosaccharide units commonly not found in mammals such as D-fucose, L-arabinose, L-rhamnose and D-galacturonic acid as well as unusual linkages of L-fucoses and sialic acids are likely the result of erroneous assigned databank entries. Microorganisms that live in mammals express a much broader variety of carbohydrate moieties and linkages and may be the source of the additional sugars.

Evaluating the accessibility of the different classes reveals that for each class even fewer building blocks are required to reach a certain number of structures. Glycolipids in comparison show a greater variety of different linkages than N- and O-linked glycans (Figure 4).

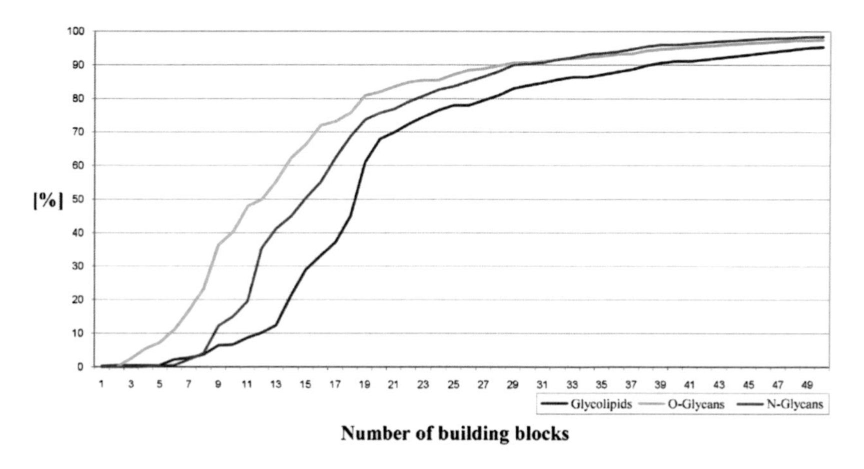

Figure 4. Percentage of accessible mammalian carbohydrates split into different classes (glycolipids, N- and O-linked glycans) and correlated to the number of building blocks.

A rather small number of building blocks is sufficient to access the majority of the mammalian glycospace. In many cases the reducing terminal units can be introduced by building blocks containing a temporary protecting group to further reduce this number. However, in the case of branched structures such an approach may be problematic. Therefore, our synthetic strategy is based on general principles with special capping building blocks.

Conclusions and Outlook

This review describes mammalian oligosaccharide diversity ("glycospace") based on a glycan databank. Carbohydrate sizes, chain lengths and branching complexity were examined. Analysis of monosaccharide connectivities within the oligomeric structures guided us to identify a set of putative monosaccharide building blocks suitable for the linear solution- and solid-phase assembly of mammalian oligosaccharides. This potential building block set was correlated with the accessible 3299 mammalian carbohydrate structures in the GLYCOSCIENCES.de databank. Only 36 building blocks are needed to construct 75% of the 3299 mammalian oligosaccharides.

Acknowledgements

I am grateful to all present and past members of my research group as well as our collaborators at a variety of institutions. I thank the Max-Planck Society for generous support.

References

[1] Seeberger, P.H, Werz, D.B. (2005) Automated Synthesis of Oligosaccharides as a Basis for Drug Discovery. *Nat. Rev. Drug Discovery* **4**:751 – 763.
doi: http://dx.doi.org/10.1038/nrd1823.

[2] a) Schmidt, R.R. (1986) New methods of glycoside and oligosaccharide syntheses – are there alternatives to the Koenigs-Knorr method? *Angew. Chem. Int. Ed. Engl.* **25**:212 – 235.
doi: http://dx.doi.org/10.1002/anie.198602121.
b) Laine, R.A. (1994) A calculation of all possible oligosaccharide isomers both branched and linear yields 1.05×10^{12} structures for a reducing hexasaccharide: the isomer barrier to development of single-method saccharide sequencing or synthesis systems. *Glycobiol*ogy **4**:759 – 767.
doi: http://dx.doi.org/10.1093/glycob/4.6.759.

[3] a) Caruthers, M.H. (1985) Gene synthesis machines: DNA chemistry and its uses, *Science* **230**:281 – 285.
doi: http://dx.doi.org/10.1126/science.3863253.
b) Caruthers, M.H. (1991) Chemical synthesis of DNA and DNA analogs. *Acc. Chem. Res.* **24**:278 – 284.
doi: http://dx.doi.org/10.1021/ar00009a005.

[4] Atherton, E., Sheppard, R.C. (1989) *Solid-Phase Peptide Synthesis: A Practical Approach*. Oxford Univ. Press, Oxford.

[5] Sears, P., Wong, C.-H. (2001) Toward automated synthesis of oligosaccharides and glycoproteins. *Science* **291**:2344 – 2350.
doi: http://dx.doi.org/10.1126/science.1058899.

[6] a) Plante, O.J., Palmacci, E.R., Seeberger, P.H. (2001) Automated solid-phase synthesis of oligosaccharides, *Science* **291**:1523 – 1527.
doi: http://dx.doi.org/10.1126/science.1057324.
b) Love, K.R., Seeberger, P.H. (2004) Automated solid-phase synthesis of protected tumor-associated antigen and blood group determinant oligosaccharides. *Angew. Chem. Int. Ed.* **43**:602 – 605.
doi: http://dx.doi.org/10.1002/anie.200352539.

c) Werz, D.B., Castagner, B., Seeberger, P.H. (2007) Automated synthesis of tumor-associated carbohydrate antigens Gb-3 and Globo-H: Incorporation of α-galactosidic linkages. *J. Am. Chem. Soc.* **129**:2770 – 2771.
doi: http://dx.doi.org/10.1021/ja069218x.

[7] von der Lieth, C.-W. (2004) An endorsement to create open access databases for analytical data of complex carbohydrates. *J. Carbohydr. Chem.* **23**:277 – 297.
doi: http://dx.doi.org/10.1081/CAR-200030093.

[8] a) Lutteke, T., Bohne-Lang, A., Loss, A., Goetz, T., Frank, M., von der Lieth, C.-W. (2006) GLYCOSCIENCES.de: an Internet portal to support glycomics and glycobiology research, *Glycobiology* **16**:71R-81R.
doi: http://dx.doi.org/10.1093/glycob/cwj049.
b) Less complex oligosaccharides may dominate due to experimental difficulties in sequencing larger structures. This problem is common to all the glycan databases.

[9] About 47% of the 3299 structures are derived from N-linked glycoproteins, 19% of them from O-linked glycoproteins and 17% from glycolipids. The rest is not assigned.

[10] a) Mootoo, D.R., Konradsson, P., Udodong, U., Fraser-Reid, B. (1988) "Armed" and "disarmed" n-pentenyl glycosides in saccharide couplings leading to oligosaccharides. *J. Am. Chem. Soc.* **110**:5583 – 5584.
doi: http://dx.doi.org/10.1021/ja00224a060.
b) Ye, X.S., Wong, C.-H. (2000) Anomeric reactivity-based one-pot oligosaccharide synthesis: A rapid route to oligosaccharide libraries. *J. Org. Chem.* **65**:2410 – 2431.
doi: http://dx.doi.org/10.1021/jo991558w.
c) Orgueira, H.A., Bartolozzi, A., Schell, P., Seeberger, P.H. (2002) Conformational locking of the glycosyl acceptor for stereocontrol in the key step in the synthesis of heparin. *Angew. Chem. Int. Ed.* **41**:2128 – 2131.
doi: http://dx.doi.org/10.1002/1521-3773(20020617)41:12<2128::AID-ANIE2128>3.0.CO;2-V.

[11] Boons, G.J., Ed. (1998) *Carbohydrate Chemistry.* Blackie, London, UK.

[12] Kunz, H., Harreus, A. (1982) Glycosidsynthese mit 2,3,4,6-tetra-*O*-pivaloyl-α-D-glucopyranosylbromid. *Liebigs Ann.* 41 – 48.

[13] Roussel, F., Knerr, L., Grathwohl, M., Schmidt, R.R. (2000) *O*-glycosyl trichloroacetimidates bearing Fmoc as temporary hydroxy protecting group: A new access to solid-phase oligosaccharide synthesis. *Org. Lett.* **2**:3043 – 3046.
doi: http://dx.doi.org/10.1021/ol006081l.

[14] Koeners, H.J., Verhoeven, J., van Boom, J.H. (1980) Synthesis of oligosaccharides by using levulinic ester as an hydroxyl protecting group. *Tetrahedron Lett.* **21**:381 – 382.
doi: http://dx.doi.org/10.1016/S0040-4039(01)85479-4.

[15] The presented building blocks (BBs) were successfully used:
a) **1**, **16**: Love, K.R., Seeberger, P.H. (2005) Solution syntheses of protected type II Lewis blood group oligosaccharides: Study for automated synthesis. *J. Org. Chem.* **70**:3168 – 3177.
doi: http://dx.doi.org/10.1021/jo047723b.
b) **2**: Tanaka, K., Goi, T., Fukase, K. (2005) Highly efficient sialylation towards α(2 – 3)- and α(2 – 6)-Neu5Ac-Gal synthesis: Significant 'fixed dipole effect' of N-phthalyl group on alpha-selectivity. *Synlett* **19**:2958 – 2962.
doi: http://dx.doi.org/10.1055/s-2005-921889.
c) **3**, **6**, **11**, **13**, **14**: Ref. 6b.
d) **4**, **8**, **12**: Hewitt, M.C., Seeberger, P.H. (2001) Automated solid-phase synthesis of a branched Leishmania cap tetrasaccharide. *Org. Lett.* **3**:3699 – 3702.
doi: http://dx.doi.org/10.1021/ol016631v.
e) **5**: Wu, X., Grathwohl, M., Schmidt, R.R. (2002) Efficient solid-phase synthesis of a complex, branched N-glycan hexasaccharide: Use of a novel linker and temporary-protecting-group pattern. *Angew. Chem. Int. Ed.* **41**:4489 – 4493.
doi: http://dx.doi.org/10.1002/1521-3773(20021202)41:23<4489::AID-ANIE4489>3.0.CO;2-X.
f) **7**; **10**, **15**, **17**, **18**: Kröck, L., Oberli, M., Werz, D.B., Seeberger, P.H., *unpublished results*.
g) **9**: Blatter, G., Beau, J.M., Jacquinet, J.C. (1994) The use of 2-deoxy-2-trichloroacetamido-D-glucopyranose derivatives in syntheses of oligosaccharides. *Carbohydr. Res.* **260**:189 – 202.
doi: http://dx.doi.org/10.1016/0008-6215(94)84038-5.
h) **19**: Wang, C.C., Lee, J.C., Luo, S.Y., Fan, H.F., Pai, C.L., Yang, W.C., Lu, L.D., Hung, S.C. (2002) Synthesis of biologically potent $\alpha 1 \rightarrow 2$-linked disaccharide derivatives via regioselective one-pot protection-glycosylation. *Angew. Chem. Int. Ed.* **41**:2360 – 2362.
doi: http://dx.doi.org/10.1002/1521-3773(20020703)41:13<2360::AID-ANIE2360>3.0.CO;2-R.
i) **20**: Belén Cid, M., Bonilla, J.B., Alfonso, F., Martín-Lomas, M. (2003) Synthesis of new hexosaminyl D- and L-chiro-inositols related to putative insulin mediators. *Eur. J. Org. Chem.* 3505 – 3515.

[16] Ratner, D.M., Swanson, E.R., Seeberger, P.H. (2003) Automated synthesis of a protected N-linked glycoprotein core pentasaccharide. *Org. Lett.* **5**:4717 – 4720.
doi: http://dx.doi.org/10.1021/ol035887t.

[17] Doubet, S., Albersheim, P. (1992) Carbbank. *Glycobiology* **2**:505.
doi: http://dx.doi.org/10.1093/glycob/2.6.505.

[18] Doubet, S., Bock, K., Smith, D., Darvill, A., Albersheim, P. (1989) The complex carbohydrate structure database. *Trends Biochem. Sci.* **14**:475 – 477.
doi: http://dx.doi.org/10.1016/0968-0004(89)90175-8.

[19] Bohne-Lang, A., Lang, E., Forster, T., von der Lieth, C.-W. (2001) LINUCS: Linear Notation for Unique Description of Carbohydrate Sequences. *Carbohydr. Res.* **336**:1 – 11.
doi: http://dx.doi.org/10.1016/S0008-6215(01)00230-0.

[20] a) de Lederkremer, R.M., Colli, W. (1995) Galactofuranose-containing glycoconjugates in trypanosomatids. *Glycobiology* **5**:547 – 552.
doi: http://dx.doi.org/10.1093/glycob/5.6.547.
b) Marlow, A.L., Kiessling, L.L. (2001) Improved Chemical Synthesis of UDP-Galactofuranose. *Org. Lett.* **3**:2517 – 2519.
doi: http://dx.doi.org/10.1021/ol016170d.

Glyco-Bioinformatics – *Bits 'n' Bytes of Sugars*
October 4th – 8th, 2009, Potsdam, Germany

Automated N-Glycan Composition Analysis with LC-MS/MSMS

Hannu Peltoniemi[1,*], Ilja Ritamo[2], Jarkko Räbinä[2] and Leena Valmu[2]

[1]Applied Numerics Ltd, Nuottapolku 10 A8, FI-00330 Helsinki, Finland.

[2]Finnish Red Cross Blood Service, R&D, Kivihaantie 7, FI-00310 Helsinki, Finland.

E-Mail: *hannu.peltoniemi@appliednumerics.fi

Received: 1st March 2010// Published: 10th December 2010

Abstract

Compared to proteomics the mass spectrometric glycan analysis still employs a lot of manual work and the differential glycomics can be a burden with increasing number of spectra. Our aim is to ease these tasks by using in-house developed glycomic software in combination with existing proteomics tools. The resulting workflow is targeted especially to glycan LC-MS/MSMS analytics and can be run with a minimal amount of human intervention. Here the method was applied to cell surface N-glycans from umbilical cord blood derived mono-nuclear cells. The final goal is to profile and differentiate the stem cell surface glycans which are being analysed at the Finnish Red Cross Blood Service.

Background

Traditionally mass spectrometric (MS) glycan analysis [1, 2] has mainly been performed by one-dimensional matrix assisted laser desorption ionization (MALDI) — time-of-flight (TOF) analysis, whereas in proteomics the use of liquid-chromatography (LC) coupled to electrospray (ESI)-MS has increased rapidly in recent years. For glycans there are many good reasons to favour MALDI, including simpler one-dimensional spectra, established wet lab procedures, existing software etc. The benefit of the LC is on the other hand in additional glycan separation, which permits for example the isomeric differentiation of glycans [3, 4].

Tandem mass spectrometric (MSMS) fragmentation analysis is also more feasible to perform on ESI-MS instrument, although it is performable today also on some MALDI instruments. The use of LC-MS/MSMS has been limited both by the complexity of the spectra, namely in the form of multiple different charge states and metal adducts, and by expensive instrumentation. Also, a major drawback is the lack of suitable software to ease the LC-MS/MSMS data analysis. Even though the available glycan software is limited, that is not the case with proteomics. A lot of software to analyse peptide LC-MS/MSMS data exists also as open source.

In the R&D department of the Finnish Red Cross Blood Service the focus is on the cell surface glycoconjugates from human cells aimed for cellular therapy. The MALDI-TOF glycan profiling of different stem cell classes, including embryonic [5], hematopoietic [6] and mesenchymal [7] cells, has previously been performed. Lately, more focused cell surface glycoconjugate analytics has been developed. Here, the cell surface proteins are biotinylated [8] and glycans are released from them. Within the workflow the glycans are further reduced in order to eliminate anomer peaks and permethylated in order to increase the ionizability in ESI and to ease the interpretation of fragmentation data [9]. The glycan samples are analyzed by reverse phase (RP)-nano-LC (LC Packings Dionex) coupled to LTQ Orbitrap XL (ThermoFisher Scientific) MS with an ESI ion source.

The experimental raw data can be represented as a LC-MS 2D map which has MSMS spectra embedded (Fig. 1). The aim of glycan identification is to find the set of glycans that *explains* the data *best*, and to calculate the total intensity (profile) for the identified glycans. Prior to the glycan identification additional pre-processing steps, for example peak picking, deisotoping and feature detection, are required.

The existing glycomic software was reviewed, but no single software to solve the overall problem was found. Among the most prominent glycan software were GlycoWorkbench [10] and Glyco-Peakfinder [11]. However, they have been developed for the analysis of one spectrum at a time and cannot be automated for a larger set of spectra acquired by LC-MS/MSMS analysis. Also, at the starting time of this project (summer 2008) the software was not yet published as open source. Other available glycan mass spectrometry related software includes free GlycoMod web tool [12] and proprietary SimGlycan [13].

For peptide LC-MS analysis there exists plenty of software both as open source (for example msInspect [14] and OpenMS [15]) and as proprietary ones (for example Progenesis LC-MS by Nonlinear Dynamics Ltd [16] and DeCyder by GE Healthcare Ltd [17]). Peptide LC-MS analysis differs slightly from the corresponding glycan analysis, mainly in the form of several charge carriers, not just hydrogen. But there are also many similar attributes in the analysis of these two analytes, peptides and glycans, including feature detection, alignment and feature comparison between different samples.

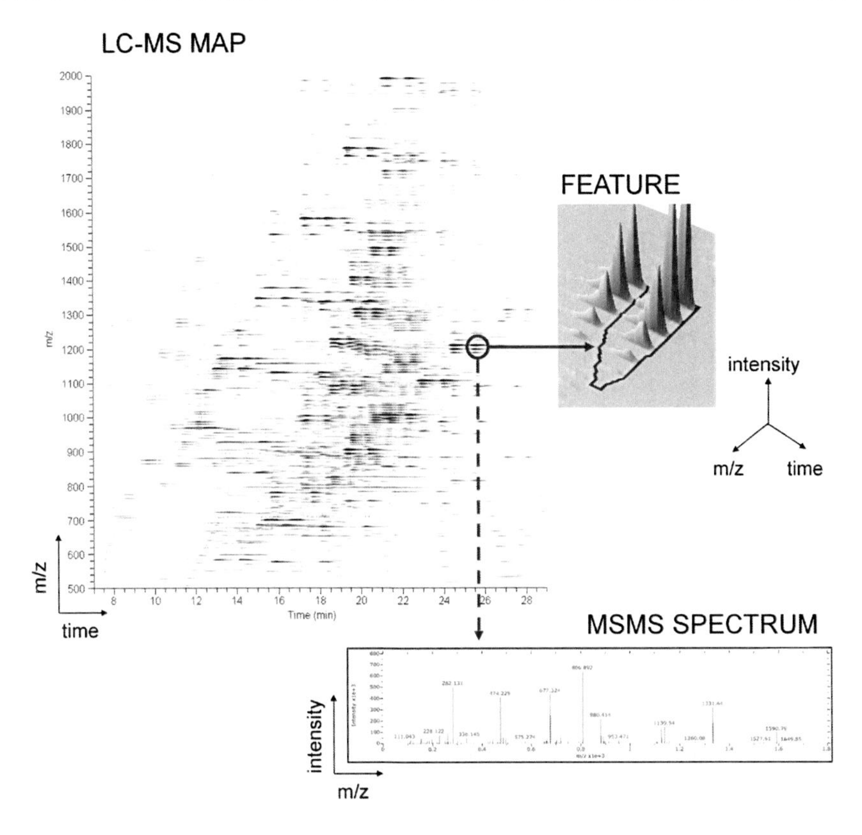

Figure 1. The LC-MS/MSMS experiment data represented as LC-MS map of eluted MS features and MSMS spectra generated by CID fragmentation of glycans.

After the survey on glycoinformatic tools the conclusion was to tailor our own glycomic software, but to apply as much of the existing software as possible, especially the tools developed for proteomics, as part of the workflow. The emphasis was more on automation and less on interactive use and user interfaces. The identification of glycan compositions was set to be sufficient to start with. To enable rapid prototyping the in-house developed part of the software was decided to be done mostly with the R statistical computing environment [18]. The R is open source software containing a lot of numerical and statistical methods, including bioinformatics methods [19].

The Glycan Identification Workflow

The glycan LC-MS/MSMS identification workflow (Fig. 2) combines existing proteomic software and in-house developed glycan specific tools. The analysis starts from a 2D LC-MS map containing chromatographic (retention time) and mass (m/z) dimensions together with embedded tandem MSMS spectra created by fragmentation of glycans. The result is a profile of matching glycans and a suggestion of the simplest set of glycans that can explain the measured data. The glycan specific tools (steps 3 – 6 in the workflow) are based on the in-house developed R library called *Glycan ID*. The library methods include spectrum matching, outlier removal, statistical scoring and visualization. The aim of the library is to enable fast development of new workflow variants when new requirements appear.

The tools in the workflow are:

1. Identify Features

Potential glycan features are identified with Progenesis LC-MS (Nonlinear Dynamics Ltd) [16] software developed originally for peptide analysis.

2. Extract MSMS Spectra

MSMS spectra with identified charge states (deisotoped) is extracted with Mascot Distiller (Matrix Science Ltd) [20]. The software has originally been developed as a pre-processor for protein identification search engine Mascot (Matrix Science Ltd).

3. Match Compositions (MS)

The glycan compositions which match to feature masses are searched. The feature matching is done either against theoretical compositions generated *de novo* with a given set of rules [21], or against a user given list of glycan compositions (database). Several charge carrier ion types and neutral adducts can be used. Outliers can be removed by iteratively applying linear fitting and elimination of compositions with a mass difference greater than two standard deviations. The tool uses an approach which is very similar to the one used by Glyco-Peakfinder [11] or GlycoMod [12].

4. Match Compositions (MSMS)

Glycan compositions which match the precursor masses and MSMS fragment spectra are searched. The precursor compositions are found as above. Fragment matching is done either against all theoretical fragments that any glycan structure with a given composition could produce [21, 22], or against theoretical or measured spectra in a given MSMS spectrum database. Outlier matches can be removed as above. The matched compositions are ranked by a statistical score defined by a logarithm of a product of two probabilities:

1) The probability that a random set of fragments would have as many or more shared peaks with the measured spectrum as the ranked composition [21] and, 2) The probability that by randomly selecting the observed number of shared peaks the same or higher amount of intensity can be covered. Two optional filtering steps are included: 1) An MSMS spectrum is taken into account only if any mass difference between two peaks matches a list of given masses, typically composed by one or two monosaccharide masses. 2) To ease the differentiation between N-acetyl-neuraminic acid (Neu5Ac) and N-glycolyl-neuraminic acid (Neu5Gc), a given residue is allowed to exist in a proposed composition only if the MSMS spectrum contains at least one of the given marker ions derived from these sialic acids.

5. Combine MS and MSMS

The results of MS and MSMS matching are combined so that the MSMS identification is included in the specific MS feature if the mass and retention time differences between the MS feature and MSMS precursor are less than the given tolerances.

6. Deconvolute

The last fully automated step in the workflow is the calculation of the total intensity and score for each proposed glycan by summing the measured feature intensities and MSMS scores with different charge states and charge carrier types, namely metal adducts and protons. The glycans are further grouped so that the proposed compositions matching a common set of features are categorized into the same group. As these sets are independent, the analysis of one group does not have an effect on the analyses of other groups. For each group, one glycan composition is marked as most likely the correct one if there is only one composition that matches all group features and if the composition has the highest score. Otherwise the group is marked to be contradictory. The glycan profile is created from the deconvoluted data and the possible contradictory groups are manually resolved based on the biological information available.

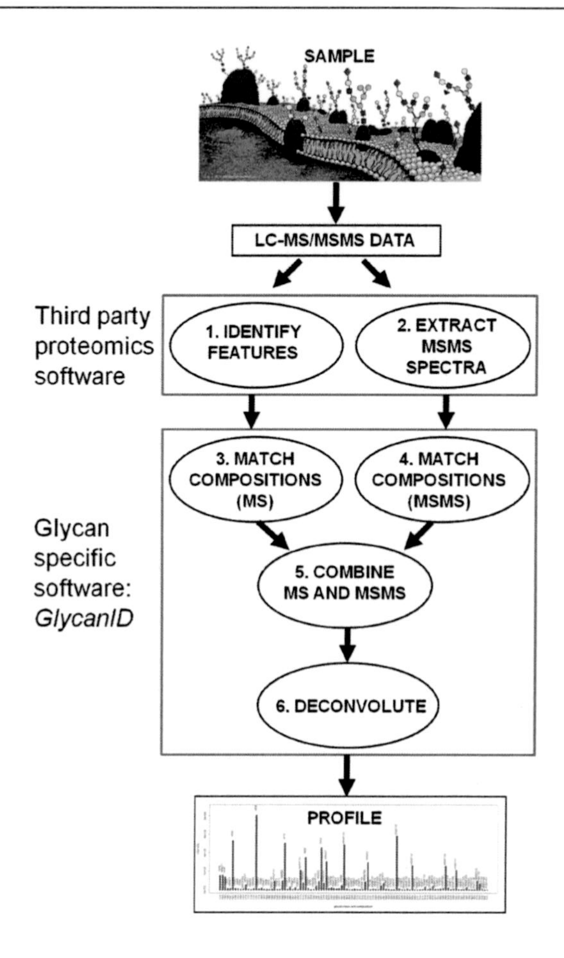

Figure 2. The glycan identification workflow.

Example: N-glycans from Umbilical Cord Blood Mononuclear Cells

The workflow was applied to the cell surface N-glycans from umbilical cord blood derived mononuclear cells. The total cellular N-glycome from the same cell type has previously been analysed by MALDI-TOF [6].

The cell surface proteins were labelled with biotin and enriched by streptavidin coupled magnetic beads as previously described [23]. N-glycans were released from the cell surface protein fraction by PNGase F and reduced with NaBH4. The reduced N-glycans were permethylated as described in [24]. The permethylated and reduced N-glycans were loaded into RP precolumn (Atlantis dC 18, Waters) and separated in analytical RP column (PepMap 100, Dionex Corporation). Ultimate 3000 LC instrument (Dionex Corporation) was operated in nano scale with a flow rate of 0.3 µl/min. The eluted glycans were introduced to LTQ Orbitrap XL mass spectrometer (Thermo Fisher Scientific Inc.) via ESI Chip interface (Advion BioSciences Inc.) in the positive-ion mode.

The glycan profiling was performed against *de novo* generated glycan compositions allowing the following restrictions: 1) Monosaccharides with 3 – 15 hexoses (H), 2 – 15 N-acetylhexosamines (N), 0 – 6 deoxy-hexoses (F) and 0 – 6 N-acetyl-neuraminic acids (S), 2) Charge carriers as either sodium or hydrogen adducts and 3) Assumption of the intact N-glycan core. Mass tolerance was set to 5 ppm with MS and to 10 ppm with MSMS spectra. The workflow started with approximately 1000 MS features and 700 MSMS precursors and ended with 54 different glycan compositions proposed by automated identification and further classified manually as biologically credible ones. About 40% of all the features had at least one matching composition, whereas the number was 90% for the 1/10 of the highest intensity features with a charge state two or higher. Naturally, the match coverage would be higher if the number of accepted monosaccharide residues and metal adducts had been larger, but also the probability to get false interpretations would increase. The future challenge will be to tune the analysis so that both the sensitivity and selectivity are optimized.

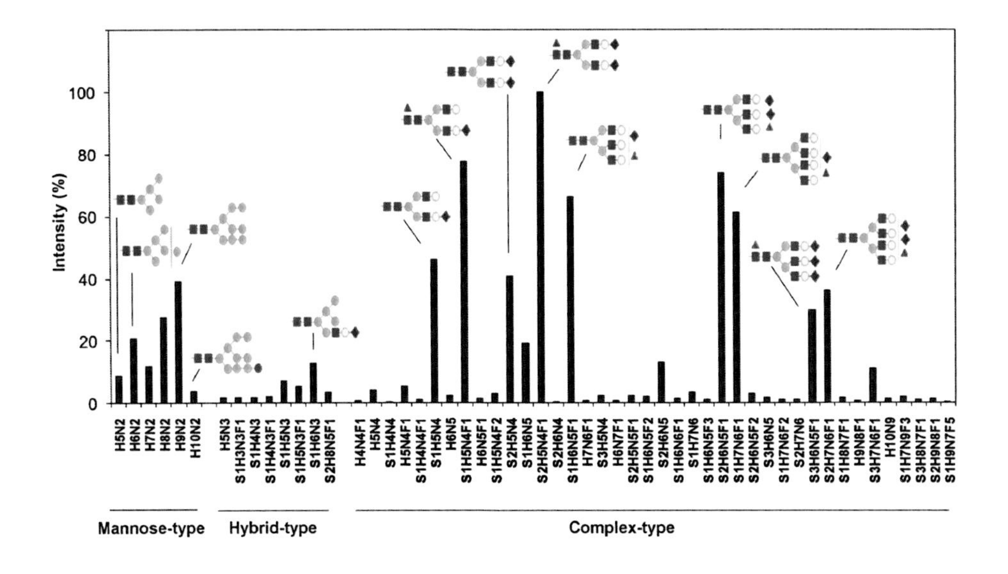

Figure 3. Cell surface N-glycan composition profile of umbilical cord blood derived mononuclear cells. The added glycan structures are based on educated guess of possible structures matching the identified compositions.

The calculated N-glycan compositions (Fig. 3) fit very well into our previously published glycan structure data of cord blood derived mononuclear cells [6], but clear differences in the cell surface N-glycan profile are seen in comparison with the total cellular N-glycan profile. On the cell surface far less high mannose-type glycans are observed, whereas complex-type glycans seem to predominate. Also fucosylated and sialylated structures are heavily enriched on the cell surface N-glycans.

Conclusions

In the mass spectrometric glycan analysis the number of features using two-dimensional LC-MS methodology is far greater than in other analytical methods typically performed in one dimension. In LC-MS analysis the complexity is increased both by ESI, which produces multiple charged ions, as well as by second dimension, introduced by chromatographic retention time. Additional complexity is still involved in the number of different metal adducts detected in the glycan analyses. Therefore, in order to utilize the additional possibilities that LC-MS analysis introduces to the glycan structure determination, a competent data handling tool is essential in order to simplify the otherwise extremely laborious interpretation of the data. However, by the limited number of monosaccharide residues allowed in the calculation and by the overlapping composition masses, some level of non-uniqueness will always be present within the given result. If an unambiguous glycan composition is required, a manual verification by an expert in glycobiology is definitely needed. The advantage of the automated glycan identification software is that it can easily show the possibilities and can generate a suggestion of a simple solution.

The glyco-bioinformatics is an emerging branch of informatics with some developed software but still with many application areas to be covered. The development of novel and automated applications could speed up if the existing software could be used as part of a novel workflow. To enable this, the software should be developed so that it can be run without a user interface, as a batch process, a library or a web service. Naturally, open source software would be the most beneficial, but proprietary ones are not excluded assuming there are no other obstacles with the workflow use. When the software presented in this study is matured enough for publication it is planned to be opened for wider use either as a web service or as open source software.

Abbreviations

ESI	Electrospray ionisation
F	deoxyhexose (fucose)
H	Hexose
LC	Liquid chromatography
MALDI	Matrix-assisted laser desorption ionisation
MS	Mass Spectrometry
MSMS	Tandem Mass Spectrometry
N	N-acetyl hexosamine
Neu5Ac	N-acetyl-neuraminic acid
Neu5Gc	N-glycolyl-neuraminic acid
RP	Reversed phase
S	Neu5Ac (sialic acid)
TOF	Time-of-flight

References

[1] North, S.J., Hitchen, P.G., Haslam, S.M., Dell, A. (2009) Mass spectrometry in the analysis of N-linked and O-linked glycans. *Curr. Opin. Struct. Biol.* **19**:498 – 506. doi: http://dx.doi.org/10.1016/j.sbi.2009.05.005.

[2] Zaia, J. (2008) Mass spectrometry and the Emerging Field of Glycomics. *Chem. Biol.* **15**:881 – 892.
doi: http://dx.doi.org/10.1016/j.chembiol.2008.07.016.

[3] Wuhrer, M., Deelder, A.M., Hokke, C.H. (2005) Protein glycosylation analysis by liquid chromatography – mass spectrometry. *J. Chrom. B* **825**:124 – 133.
doi: http://dx.doi.org/10.1016/j.jchromb.2005.01.030.

[4] Ruhaak, L.R., Deelder, A.M., Wuhrer, M. (2009) Oligosaccharide analysis by graphitized carbon liquid chromatography – mass spectrometry. *Anal. Bioanal. Chem.* **394**:163 – 174.
doi: http://dx.doi.org/10.1007/s00216-009-2664-5.

[5] Satomaa, T., Heiskanen, A., Mikkola, M., Olsson, C., Blomqvist, M., Tiittanen, M., Jaatinen, T., Aitio, O., Olonen, A., Helin, J., Hiltunen, J., Natunen, J., Tuuri, T., Otonkoski, T., Saarinen, J., Laine, J. (2009) The N-glycome of human embryonic stem cells. *BMC Cell Biol.* **10**:42.
doi: http://dx.doi.org/10.1186/1471-2121-10-42.

[6] Hemmoranta, H., Satomaa, T., Blomqvist, M., Heiskanen, A., Aitio, O., Saarinen, J., Natunen, J., Partanen, J., Laine, J., Jaatinen, T. (2007) N-glycan structures and associated gene expression reflect the characteristic N-glycosylation pattern of human hematopoietic stem and progenitor cells. *Exp. Hematol.* **35**:1279 – 1292.
doi: http://dx.doi.org/10.1016/j.exphem.2007.05.006.

[7] Heiskanen, A., Hirvonen, T., Salo, H., Impola, U., Olonen, A., Laitinen, A., Tiitinen, S., Natunen, S., Aitio, O., Miller-Podraza, H., Wuhrer, M., Deelder, A.M., Natunen, J., Laine, J., Lehenkari, P., Saarinen, J., Satomaa, T., Valmu, L. (2009) Glycomics of bone marrow-derived mesenchymal stem cells can be used to evaluate their cellular differentiation stage. *Glycoconj. J.* **26**:367 – 384.
doi: http://dx.doi.org/10.1007/s10719-008-9217-6.

[8] Elia, G. (2008) Biotinylation reagents for the study of cell surface proteins. *Proteomics* **8**:4012 – 4024.
doi: http://dx.doi.org/10.1002/pmic.200800097.

[9] Costello, C., Contado-Millera, J.M., Cipollo, J.F. (2007) A glycomics platform for the analysis of permethylated oligosaccharide alditols. *J. Am. Soc. Mass Spectrom.* **18**:1799 – 1812.
doi: http://dx.doi.org/10.1002/pmic.200800097.

[10] Ceroni, A., Maass, K., Geyer, H., Geyer, R., Dell, A., Haslam, S.M. (2008) GlycoWorkbench: A Tool for the Computer-Assisted Annotation of Mass Spectra of Glycans. *J. Proteome Res.* **7**:1650 – 1659.
doi: http://dx.doi.org/10.1002/pmic.200800097.

[11] Maass, K., Ranzinger, R., Geyer, H., von der Lieth, C-W., Geyer, R. (2007) "GlycoPeakfinder" – *de novo* composition analysis of glycoconjugates. *Proteomics* **7**:4435 – 4444.
doi: http://dx.doi.org/10.1002/pmic.200700253.

[12] Cooper, C.A., Gasteiger, E., Packer, N.H. (2001) GlycoMod – A software Tool for Determining Glycosylation Compositions from Mass Spectrometric Data. *Proteomics* **1**:340 – 349.
doi: http://dx.doi.org/10.1002/1615-9861(200102)1:2<340::AID-PROT340>3.3.CO;2-2.

[13] SimGlycan http://www.premierbiosoft.com/glycan/index.html

[14] May, D., Law, W., Fitzgibbon, M., Fang, Q., McIntosh, M. (2009) Software Platform for Rapidly Creating Computational Tools for Mass Spectrometry-Based Proteomics. *J. Proteome Res.* **8**:3212 – 3217. doi: http://dx.doi.org/10.1021/pr900169w.

[15] Kohlbacher, O., Reinert, K. (2009) OpenMS and TOPP: Open Source Software for LC-MS Data Analysis. In: *Proteome Bioinformatics*, ed. by Simon J. Hubbard and Andrew R. Jones, Methods in Molecular Biology. Humana Press., vol. 604, chap. 14.

[16] Progenesis LC-MS, http://www.nonlinear.com

[17] DeCyder MS Differential Analysis Software, http://www.gelifesciences.com

[18] The R Project for Statistical Computing, http://www.r-project.org.

[19] BioConductor, http://www.bioconductor.org.

[20] Mascot Distiller, http://www.matrixscience.com

[21] Joenväärä, S., Ritamo, I., Peltoniemi, H., Renkonen, R. (2008) N-Glycoproteomics – an automated workflow approach. *Glycobiology* **18**:339 – 349.

[22] Peltoniemi, H., Joenväärä, S., Renkonen, R. (2009) De novo glycan structure search with the CID MS/MS spectra of native N-glycopeptides. *Glycobiology* **19**: 707 – 714.

[23] Scheurer, S.B., Rybak, J-N., Roesli, C., Brunisholz, R.A., Potthast, F., Schlapbach, R., Neri, D., Elia, G. (2005) Identification and relative quantification of membrane proteins by surface biotinylation and two-dimensional peptide mapping. *Proteomics* **5**:2718 – 2728.

[24] Kang, P., Mechref, Y., Klouckova, I., Novotny, M.V. (2005) Solid-phase permethylation of glycans for mass spectro-metric analysis. *Rapid Commun. Mass Spectrom.* **19**:3421 – 3428.

Glyco-Bioinformatics – *Bits 'n' Bytes of Sugars*
October 4th – 8th, 2009, Potsdam, Germany

Glycoinformatic Platforms for Data Interpretation: An HPLC Perspective

Matthew P. Campbell*, Natalia V. Artemenko and Pauline M. Rudd

NIBRT Dublin-Oxford Glycobiology Laboratory, National Institute for Bioprocessing Research and Training, Conway Institute, University College Dublin, Ireland

E-Mail: *matthew.campbell@nibrt.ie

Received: 1st March 2010 / Published: 10th December 2010

Abstract

High-throughput and automated HPLC techniques allow for the rapid, detailed structural analysis of complex glycans. These advances have the potential for (i) validating a new generation of biomarkers that relate alterations in glycan processing to disease by mining the glycosylation patterns in a variety of disease types and (ii) for monitoring the production of therapeutic glycoproteins. The wealth of knowledge that can be generated justifies the requirement for databases, analytical tools and search facilities. Recent efforts by international consortia have increased the awareness of the need for glycoinformatics and several resources are now available including a suite of novel applications to assist the interpretation of HPLC data collections.

Introduction

Glycosylation is the most common and structurally diverse post-translational modification of proteins that has an impact on a wide range of biological functions [1]. It can have profound effects on the structure of and physico-chemical properties of proteins and their activity to half of all cellular and secretory proteins [2]. It is widely accepted that carbohydrates are important factors in many biological recognition processes and our understanding of their functions is rapidly expanding with advances in high-throughput glycomic strategies, disease/biomarker profiling, improved understanding of the molecular glycosylation machinery and glycoinformatics.

The biosynthesis of glycans is a non-template-driven process in which saccharide donors, glycosyltransferases, and exoglycosidases play an interactive role. The resulting glycans can have complex structures with multiple branching points where each monosaccharide component can be one of approximately 15 residues present in nature [3]. Previous reports suggest that aberrant changes in protein glycosylation, specifically the attachment of specific monosaccharide residues or branch changes, have implications in various biological and pathological processes including infection, autoimmune disorders and cancer [4 – 7]. The characterization of disease-associated glycans from serum glycoproteins has helped our understanding of disease pathologies and provides the potential for identifying biomarkers for diagnosis and prognosis [8 – 11].

The inherent complexity of glycan structures and microheterogeneity (glycoforms in which a single protein is diversified by a heterogeneous array of glycans at each glycosylation site) makes analysis of glycoconjugates very challenging. Glycan analysis relies on the ability to detect small quantities of glycans on low abundant glycoproteins since there is a requirement to characterize glycans at the 1% level. A full and detailed characterization of glycan structure is a time consuming analytical process dependent on an array of sensitive, robust, high-resolution separation technologies that are needed to determine monosaccharide composition, linkages and branching sequences. Those currently in use include HPLC, Mass Spectrometry, tandem MS (MS/MS), LC-MS, capillary electrophoresis (CE), high-performance anion exchange chromatography with pulsed amperometric detection (HPAEC-PAD), NMR and glycan arrays [12 – 17]

High performance liquid chromatography (HPLC) and mass spectrometry (MS) are the most widely used techniques to address the challenges as they offer high levels of sensitivity and the ability to handle complex mixtures of different glycan variations. There are advantages and disadvantages to each technique and many strategies incorporate an orthogonal approach due to the complex nature of glycans, where no single method can fully characterize a given glycan structure and/or function [17, 12].

Modern high-throughput HPLC-based methods are well-established and powerful methods used to obtain quantitative, reproducible and high-resolution separations of glycans at the femtomole level. The major benefits of this technology include, firstly, a well established technique using amide-based columns to generate highly reproducible and resolved separation profiles that is capable of separating structures with the same composition on the basis of sequence and linkage type; the ability to analyze both neutral and charged glycans in the same run; the 1:1 stoichiometry labelling of glycan with the fluorophore 2-aminobenzamide (2AB) that permits quantification; the normalization of glycans identified by using a standard, for example dextran, that enables the calibration of retention times to glucose unit values (GU) using Empower GPC software from Waters Ltd; finally, the use of specific exoglyosidases for the sequential digestion of glycan pools that allows complete characterization of complex glycans [16, 18].

Previously we have reported a high-throughput 96 well platform method for N-glycan analysis [16]. Recent method development and advances to this technology have included interfacing fully automated liquid-handling robotic platforms, to optimize sample immobilization, enzymatic N-glycan release and fluorescent labelling; increasing sample productivity and reducing the time to complete sample preparation to eight hours. The availability of robotic-based instrumentation complements biomarker discovery and validation programs by accelerating the analysis of large sample cohorts where high-throughput methods are a prerequisite. Furthermore, automated platforms increase the application of glycomics by supporting alternative separation technologies, for example, specifically in the bioprocessing industry including quality by design, process analytical technology and critical feature analysis (glycosylation monitoring for therapeutic protein manufacturing) [20].

Glycoinformatics

The increasing amounts of data generated by mass spectrometry and high-throughput HPLC glycan analysis combined with high sensitivity, improved cycle speeds and robotic sample and handling platforms requires new ideas and methods for disseminating the conclusions drawn from such experiments. The sheer complexity and volume of data routinely generated necessitates bioinformatics solutions in the form of data repositories and analytical tools to facilitate data interpretation.

The traditional route of publishing articles is still the most practical method, however only a small proportion of the results and underlying data are readily accessible. When HPLC and mass spectrometry data is available it is generally only provided in specific formats. In addition the size of the data files result in raw data being archived and stored on local servers with only the final processed data being retained. To enable the community to effectively mine an increasingly rich data source, the experimental data needs to be collected in central public repositories.

In contrast to the genomic and proteomic areas, the glycosciences lack accessible, curated and comprehensive data collections that summarize the structure, characteristics, biological origin and potential function of glycans that have been experimentally verified and reported in the literature. The complexity of glycan structures and the variety of techniques available pose additional obstacles to the development of an accurate and user-friendly suite of tools and databases.

The current trend and population of databases is characterized by the existence of disconnected and incompatible collections of experimental data and proprietary applications. The lack of compatibility between these existing databases and there sparseness hinders the development of bioinformatics tools for the interpretation of data and implementing platforms for large-scale glycomics and glycoproteomic studies.

As glycan-related databases improve in both coverage and quality developers and experimentalists need to consider solutions and criteria requirements that maximize the value that can be extracted. The community has recognized the need for a more organized approach to accessing glycan related data. Over the last few years a number of international consortia have developed frameworks that support the growing requirement for storing and analyzing analytical data. Recent reviews highlight the approaches and difficulties the glycoinformatic community are facing and those steps being taken to create well-curated and annotated databases and tools [20, 21].

An Infrastructure Design Study (EUROCarbDB) was started under the sixth EU framework programme to establish the technical requirements for such a centralised and standardised architecture. The platform provides an introduction and/or recommendation of formats and nomenclatures; user-curated structural and experimental data; and open access to software programs and libraries to support continued development (http://www.eurocarbdb.org).

The partners of the EUROCarbDB initiative have developed tools and work flows to assist the interpretation of HPLC, MS and NMR experimental data, extending this infrastructure will utilize a strong background of resources to develop tools for data sharing and collection. This integration will meet the objectives set out by the ESF policy briefing (Structural Medicine – The Importance of Glycomics for Health and Disease, ESF Science Policy Briefing 27; 2006) and the NIH white paper [21]. A detailed overview of EUROCarbDB will be published elsewhere.

Recently, during a Consortium for Functional Glycomics (CFG) workshop on Analytical and Bioinformatic Glycomics a new Working Group on Glycomics Data(base) Standards (WGGDS) was established with the aim of defining standards and work flows for glycan data exchange. The WGGDS includes leading research partners from the CFG, GlycomeDB, RINGS and the EUROCarbDB consortia involved in glycomics and glycoinformatics.

The fundamental aim of EUROCarbDB and WGGDS is to make it possible for labs to run experiments and combine results, to distribute workloads, provide access to a selection of tools, and to derive new methodologies by comparing results to help realise the potential of high-throughput glycomics. Each consortium and partner member can provide an extremely useful data resource contributing data acquired by the group or retrieved from external sources.

In addition to the EUROCarbDB and CFG resources the other most prominent and publicly available databases include the Kyoto Encyclopedia of Genes and Genomes glycome portal (KEGG GLYCAN) [22], the Japanese Consortium for Glycobioloy and Glycotechnology Databases (JCGGDB) and Glycoscience.de [23]. Other efforts in recent years are: the Russian Bacterial Carbohydrate Structure Database (BCSDB) [24] and GlycomeDB [25], an initiative to retrieve all structure entries, taxonomic annotations, and references from major databases. Most recently GlycoSuite [26], originally a commercial database, is now hosted by the Swiss Bioinformatics Institute.

However, further developments are required to establish a comprehensive infrastructure required for next generation analytical and informatic tools. Especially, collaborative efforts to standardize formats for structural, analytical, and reference data including controlled vocabularies. For example, a glycomics experiment can include many different terms and descriptions, therefore, many frameworks including EUROCarbDB have develop internal vocabularies or reference well established and controlled vocabularies or ontologies, for example the NCBI taxonomy and MeSH terminologies. The availability of maintained controlled vocabularies has provided the stability required for database and software development.

HPLC Data Analysis: Developing Tools and Databases

The complexity of glycoconjugates and techniques used to elucidate their structures does present significant bottlenecks to the development of integrated software and database packages. There are an increasing number of tools and database to support glycomics investigations especially for HPLC [27, 28], mass spectrometry [29 – 31] and NMR [32].

Recently, several new HPLC-based tools have been developed in conjunction with EUROCarbDB to help populate databases and to assist the process of data analysis and annotation. These tools and work flows have been designed primarily to support the group's high-throughput strategy with an open-source philosophy that is intended to progress the application of glycoinformatics in glycomics and glycoproteomics. This is achieved by the provision of semi-automated capabilities inclusive of a robust database framework, formats for data exchange, and support for storing and retrieving glycan structural and experimental data.

An innovative suite of integrated analytical tools, databases and standards are being developed to support the growing demand for HPLC data interpretation and access to well curated data collections such as GlycoBase, autoGU [28] and GlycoExtractor [27] (http://glycobase.nibrt.ie).

GLYCOBASE AND AUTOGU

GlycoBase is a novel experimental database developed to support our HPLC methodology. The database contains the HPLC elution positions for 2AB-labelled N- and O-linked glycan structures (expressed in the form of glucose unit values), the predicted products of exoglycosidase digestions; supporting literature information; and a listing of subgroups in which the glycan has been identified. All structures were determined by a combination of Normal Phase-HPLC with exoglycosidase sequencing and mass spectrometry (MALDI-MS, ESI-MS, ESI-MS/MS, LC-MS, LC-ESI-MS/MS). Each carbohydrate structure is stored in the database using the GlycoCT [33] format that can be used to dynamically convert the pictorial representation of structures to a series of support nomenclatures including Oxford University [34] and CFG black and white and colour formats and textual representation. This is achieved by using the application programming interface created by EUROCarbDB; a detailed discussion of the EUROCarbDB features, frameworks and database structure will be published elsewhere.

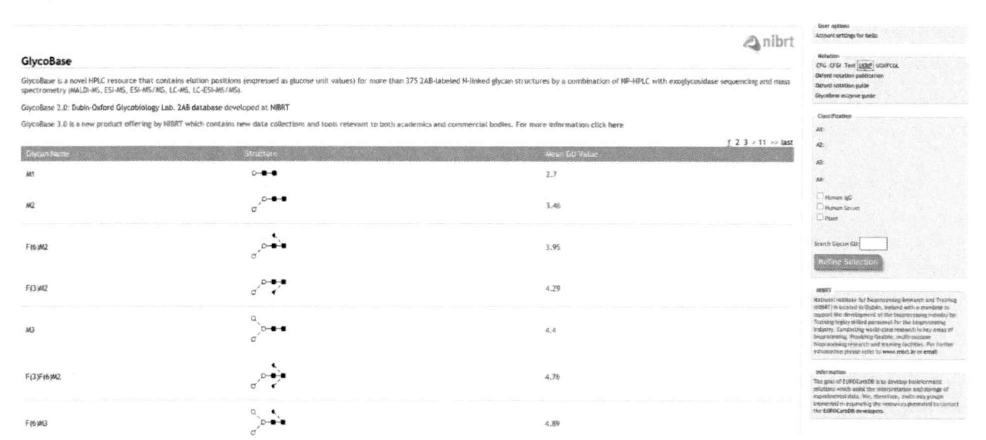

Figure 1. An overview of the front page to GlycoBase using the EUROCarbDB template. The page lists all structures stored in the database including a short name description; a pictorial representation in the Oxford nomenclature system which can be converted to other supported formats using the GlycoCT and EUROCarbDB API; and average reported glucose unit values. The refinement options on the right panel allow the user to display only those structures with specific structural features have been classified into specific subsets and/or fall within a user defined glucose unit range.

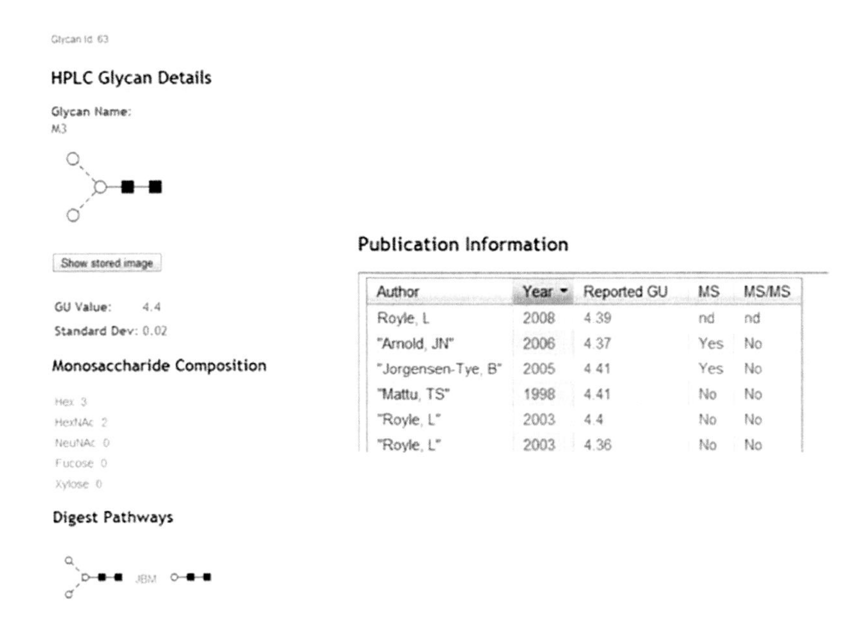

Author	Year	Reported GU	MS	MS/MS
Royle, L	2008	4.39	nd	nd
"Arnold, JN"	2006	4.37	Yes	No
"Jorgensen-Tye, B"	2005	4.41	Yes	No
"Mattu, TS"	1998	4.41	No	No
"Royle, L"	2003	4.4	No	No
"Royle, L"	2003	4.36	No	No

Figure 2. A data entry page listing for the Man3 glycan. For each glycan entry a description of the mean and standard deviation values is listed for all published references. In addition the monosaccharide composition is displayed followed by verified exoglycosidase digestions pathways. The publication of glucose unit values and products of digestions are essential aids to the manual interpretation of HPLC data collections.

The interpretation of HPLC data including exoglycosidase digestions can be time consuming and database-matching software (autoGU) is available to assist the assignment of possible glycan structures to each HPLC peak. When used in combination with data from a series of exoglycosidases, autoGU will create a refined list of structures based on the digest footprint i. e. shifts in GU values due to cleavage of terminal monosaccharides dependent on enzyme specificity.

GlycoBase and autoGU provide the backbone for the development of next generation glycoinformatics tools that are an invaluable aid to data interpretation. They provide the support for manual data interpretation whilst an active development program includes a number of new user-friendly features to meet the requirements of the end-user.

GlycoExtractor

A recent application, GlycoExtractor, is a novel approach for extracting large volumes of processed HPLC data from proprietary chromatography data software that stores locally acquired sample runs (Waters Empower). It is a web-based application that facilitates user demands to extract a series of sample sets generated by high-throughput methods. The tool was developed to improve existing methods for processing and exporting large volumes of information for use with autoGU. Current methods are cumbersome, partly due to the lack of embedded workflows for exporting multiple sample sets. GlycoExtractor automates routine tasks by querying specific data attributes across all active projects to a single file rather than a set of disconnected output files, improving data interpretation [27].

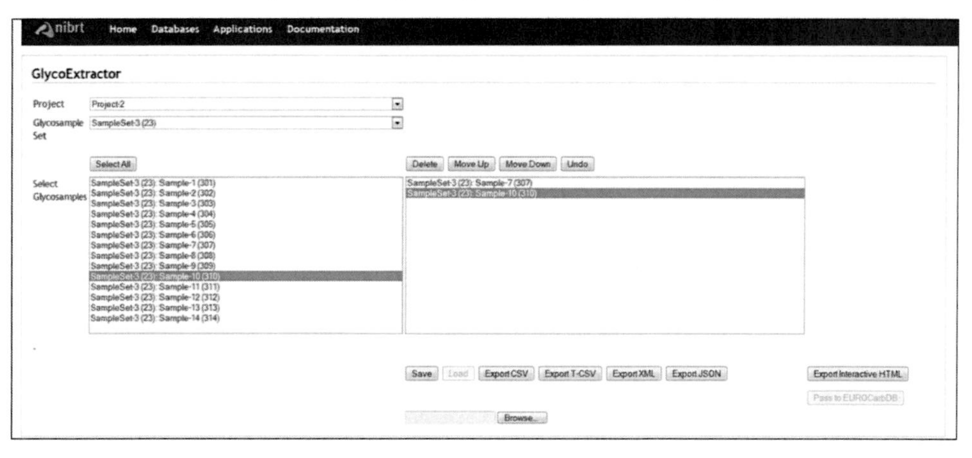

Figure 3. GlycoExtractor interfacing with an example database of HPLC data. The corresponding glucose unit values, peaks areas and project summary can be exported to a single file in XML, JSON or CSV format.

The development of these novel tools was achieved by EU efforts to develop centralized resources for the glycobiology community. It is anticipated that the next release will be a fully integrated solution of tools interfacing with the EUROCarbDB framework that will further improve both data processing and interpretation.

Conclusions

The availability and development of glycoinformatics databases and tools have increased considerably in recent years as a result of the efforts of international collaborations and consortia. However, further development is required to establish a comprehensive infrastructure required for next generation analytical and bioinformatics tools that rivals those platforms available for genomics and proteomics.

The formation of partnerships between leading consortia, such as the recent CFG sponsored WGGDS program is opening opportunities for developers and experimentalists to develop an infrastructure for cross-referencing data collections; the integration of resources that allows users to search data collections in a unified and user-friendly way. It is essential that the work achieved to date continues and the availability of organized and annotated databases of analytical data will enable the development of new technologies and supporting tools for automated, high-throughput identification of glycan structures.

Acknowledgements

The authors would like to thank the collaboration with the EUROCarbDB initiative (http://www.eurocarbdb.org) RIDS contract number: 011952.

References

[1] Dwek, R.A. (1996) Glycobiology: Toward Understanding the Function of Sugars. *Chem. Rev.* **96**: 683 – 720.
doi: http://dx.doi.org/10.1021/cr940283b.

[2] Apweiler, R., Hermjakob, H., Sharon, N. (1999) On the frequency of protein glycosylation, as deduced from analysis of the SWISS-PROT database. *Biochim. Biophys. Acta* **1473**:4 – 8.

[3] Helenius, A., Aebi, M. (2004) Roles of N-linked glycans in the endoplasmic reticulum. *Annu. Rev. Biochem.* **73**:1019 – 1049.
doi: http://dx.doi.org/10.1146/annurev.biochem.73.011303.073752.

[4] Fuster, M.M., Esko, J.D. (2005) The sweet and sour of cancer: glycans as novel therapeutic targets. *Nat. Rev. Cancer* **5**:526 – 542.
doi: http://dx.doi.org/10.1038/nrc1649.

[5] Green, R.S., Stone, E.L., Tenno, M., Lehtonen, E., *et al.* (2007) Mammalian N-glycan branching protects against innate immune self-recognition and inflammation in autoimmune disease pathogenesis. *Immunity* **27**:308 – 320.
doi: http://dx.doi.org/10.1016/j.immuni.2007.06.008.

[6] Hakomori, S. (2002) Glycosylation defining cancer malignancy: new wine in an old bottle. *Proc. Natl. Acad. Sci. U.S.A.* **99**:10231 – 10233.
doi: http://dx.doi.org/10.1073/pnas.172380699.

[7] Taniguchi, N., Ekuni, A., Ko, J.H., Miyoshi, E., *et al.* (2001) A glycomic approach to the identification and characterization of glycoprotein function in cells transfected with glycosyltransferase genes. *Proteomics* **1**:239 – 247.
doi: http://dx.doi.org/10.1002/1615-9861(200102)1:2<239::AID-PROT239>3.0.CO;2-K.

[8] Abd Hamid, U.M., Royle, L., Saldova, R., Radcliffe, C.M., *et al.* (2008) A strategy to reveal potential glycan markers from serum glycoproteins associated with breast cancer progression. *Glycobiology* **18**:1105 – 1118.
doi: http://dx.doi.org/10.1093/glycob/cwn095.

[9] Arnold, J.N., Saldova, R., Hamid, U.M.A., Rudd, P.M. (2008) Evaluation of the serum N-linked glycome for the diagnosis of cancer and chronic inflammation. *Proteomics* **8**:3284 – 3293.
doi: http://dx.doi.org/10.1002/pmic.200800163.

[10] Butler, M., Quelhas, D., Critchley, A.J., Carchon, H., *et al.* (2003) Detailed glycan analysis of serum glycoproteins of patients with congenital disorders of glycosylation indicates the specific defective glycan processing step and provides an insight into pathogenesis. *Glycobiology* **13**:601 – 622.
doi: http://dx.doi.org/10.1093/glycob/cwg079.

[11] Wuhrer, M. (2007) Glycosylation profiling in clinical proteomics: heading for glycan biomarkers. *Expert. Rev. Proteomics* **4**:135 – 136.
doi: http://dx.doi.org/10.1586/14789450.4.2.135.

[12] Domann, P.J., Pardos-Pardos, A.C., Fernandes, D.L., Spencer, D.I.R., *et al.* (2007) Separation-based glycoprofiling approaches using fluorescent labels. *Proteomics* **7**(Suppl. 1):70 – 76.
doi: http://dx.doi.org/10.1002/pmic.200700640.

[13] Estrella, R.P., Whitelock, J.M., Roubin, R.H., Packer, N.H., Karlsson, N.G. (2009) Small-scale enzymatic digestion of glycoproteins and proteoglycans for analysis of oligosaccharides by LC-MS and FACE gel electrophoresis. *Methods Mol. Biol.* **534**:171 – 192.

[14] Harvey, D.J. (2005) Proteomic analysis of glycosylation: structural determination of N- and O-linked glycans by mass spectrometry. *Expert. Rev. Proteomics* **2**:87 – 101.
doi: http://dx.doi.org/10.1586/14789450.2.1.87.

[15] Harvey, D.J. (2009) Analysis of carbohydrates and glycoconjugates by matrix-assisted laser desorption/ionization mass spectrometry: An update for 2003 – 2004. *Mass Spectrom. Rev.* 28:273 – 361.
doi: http://dx.doi.org/10.1002/mas.20192.

[16] Royle, L., Campbell, M.P., Radcliffe, C.M., White, D.M., *et al.* (2008) HPLC-based analysis of serum N-glycans on a 96-well plate platform with dedicated database software. *Ana.l Biochem.* **376**:1 – 12.
doi: http://dx.doi.org/10.1016/j.ab.2007.12.012.

[17] Wada, Y., Azadi, P., Costello, C.E., Dell, A., *et al.* (2007) Comparison of the methods for profiling glycoprotein glycans – HUPO Human Disease Glycomics/Proteome Initiative multi-institutional study. *Glycobiology* **17**:411 – 422.
doi: http://dx.doi.org/10.1093/glycob/cwl086.

[18] Royle, L., Dwek, R.A., Rudd, P.M. (2006) Determining the structure of oligosaccharides N- and O-linked to glycoproteins. *Curr. Protoc. Protein Sci. Chapter 12* Unit 12.16.

[19] Blow, N. (2009) Glycobiology: A spoonful of sugar. *Nature* **457**:617 – 620. doi: http://dx.doi.org/10.1038/457617a.

[20] Aoki-Kinoshita, K.F. (2008) An introduction to bioinformatics for glycomics research. *PLoS Comput. Biol.* **4**:e1000075. doi: http://dx.doi.org/10.1371/journal.pcbi.1000075.

[21] Packer, N.H., von der Lieth, C.-W., Aoki-Kinoshita, K.F., Lebrilla, C.B., *et al.* (2008) Macquarie University, Sydney, New South Wales, NSW 2109, Australia. 2008, pp. 8 – 20.

[22] Hashimoto, K., Goto, S., Kawano, S., Aoki-Kinoshita, K.F., *et al.* (2006) KEGG as a glycome informatics resource. *Glycobiology* **16**:63R-70R. doi: http://dx.doi.org/10.1093/glycob/cwj010.

[23] Lutteke, T., Bohne-Lang, A., Loss, A., Goetz, T., *et al.* (2006) GLYCOSCIENCES.de: an internet portal to support glycomics and glycobiology research. *Glycobiology* **16**:71R-81R. doi: http://dx.doi.org/10.1093/glycob/cwj049.

[24] Toukach, P., Joshi, H. J., Ranzinger, R., Knirel, Y., von der Lieth, C.-W. (2007) Sharing of worldwide distributed carbohydrate-related digital resources: online connection of the Bacterial Carbohydrate Structure DataBase and GLYCOSCIENCES.de. *Nucleic Acids Res.* **35**:D280 – 286. doi: http://dx.doi.org/10.1093/nar/gkl883.

[25] Ranzinger, R., Frank, M., von der Lieth, C.-W., Herget, S. (2009) Glycome-DB.org: a portal for querying across the digital world of carbohydrate sequences. *Glycobiology* **19**:1563 – 1567. doi: http://dx.doi.org/10.1093/glycob/cwp137.

[26] Cooper, C.A., Harrison, M.J., Wilkins, M.R., Packer, N.H. (2001) GlycoSuiteDB: a new curated relational database of glycoprotein glycan structures and their biological sources. *Nucleic Acids Res.* **29**:332 – 335. doi: http://dx.doi.org/10.1093/nar/29.1.332.

[27] Artemenko, N.V., Campbell, M.P., Rudd, P.M. (2010) GlycoExtractor – a web-based interface for high throughput processing of HPLC-glycan data. *J. Proteome Res.* **9**(4):2037 – 2041. doi: http://dx.doi.org/10.1021/pr901213u.

[28] Campbell, M.P., Royle, L., Radcliffe, C.M., Dwek, R.A., Rudd, P.M. (2008) GlycoBase and autoGU: tools for HPLC-based glycan analysis. *Bioinformatics* **24**:1214 – 1216.
doi: http://dx.doi.org/10.1093/bioinformatics/btn090.

[29] Goldberg, D., Sutton-Smith, M., Paulson, J., Dell, A. (2005) Automatic annotation of matrix-assisted laser desorption/ionization N-glycan spectra. *Proteomics* **5**:865 – 875.
doi: http://dx.doi.org/10.1002/pmic.200401071.

[30] Ceroni, A., Maass, K., Geyer, H., Geyer, R., *et al.* (2008) GlycoWorkbench: a tool for the computer-assisted annotation of mass spectra of glycans. *J. Proteome Res.* **7**:1650 – 1659.
doi: http://dx.doi.org/10.1021/pr7008252.

[31] Maass, K., Ranzinger, R., Geyer, H., von der Lieth, C.W., Geyer, R. (2007) "Glycopeakfinder" – *de novo* composition analysis of glycoconjugates. *Proteomics* **7**:4435 – 4444.
doi: http://dx.doi.org/10.1002/pmic.200700253.

[32] Loss, A., Stenutz, R., Schwarzer, E., von der Lieth, C.W. (2006) GlyNest and CASPER: two independent approaches to estimate 1 H and 13C NMR shifts of glycans available through a common web-interface. *Nucleic Acids Res.* **34**:W733 – 737.
doi: http://dx.doi.org/10.1093/nar/gkl265.

[33] Herget, S., Ranzinger, R., Maass, K., Lieth, C.W. (2008) GlycoCT-a unifying sequence format for carbohydrates. *Carbohydr. Res.* **343**:2162 – 2171.
doi: http://dx.doi.org/10.1016/j.carres.2008.03.011.

[34] Harvey, D.J., Merry, A.H., Royle, L., Campbell, M.P., *et al.* (2009) Proposal for a standard system for drawing structural diagrams of N- and O-linked carbohydrates and related compounds. *Proteomics* **9**:3796 – 3801.
doi: http://dx.doi.org/10.1002/pmic.200900096.

Glyco-Bioinformatics – *Bits 'n' Bytes of Sugars*
October 4th – 8th, 2009, Potsdam, Germany

ProSim: Development of a User-friendly Molecular Modelling Package

Hans Heindl[1], Tamas Kiss[2], Gabor Terstyanszky[2], Noam Weingarten[2], Babak Afrough[3] and Pamela Greenwell[1,*]

[1]School of Life Sciences, University of Westminster, London, U.K.

[2]Centre for Parallel Computing, School of Electronics and Computing, University of Westminster, London, U.K.

[3]Institute for Women's Health, University College London, London, U.K.

E-Mail: *greenwp@wmin.ac.uk

Received: 18th March 2010 / Published: 10th December 2010

Abstract

We have developed and tested a user-friendly automated molecular modelling workflow with a web-based interface. The workflow has been tested using protein: drug, enzyme: substrate and lectin: sugar interactions. The work presented here involves studies using 3 glycosidases (a human mannosidase, a viral neuraminidase and a protozoan sialidase). We have illustrated the utility of the workflow using the mannosidase and thio-saccharide inhibitor. We have shown the limitations of such *in silico* technology when working with enzymes like the viral neuraminidase in which dynamic conformation changes take place during the binding or where access to the active site is blocked by a loop or a single residue. The viral neuraminidase inhibitor does not bind to the protozoan sialidase suggesting that such inhibitors would have no use in therapeutic strategies.

Background

The work to be described involved the development and validation of a web-interfaced workflow incorporating energy minimisation, docking and molecular dynamics tools for the *in silico* modelling of proteins and their ligands. This worked formed part of a JISC supported project, ProSim, the objectives of which were:

1. To define user requirements and user scenarios of the protein molecule simulation involving energy minimization, checking and validating 3D structures, docking ligand to receptor molecule and refining/*de novo* prediction of ligand-receptor molecule interactions
2. To identify software packages required for protein molecule simulation, test these packages and select those suitable.
3. To automate the protein molecule simulation creating workflow templates and to provide parameter study support.
4. To develop application specific graphical user interfaces (application portlets) to provide seamless access to Grid resources and services.
5. To port and run protein molecule simulation on the National Grid Service (NGS) and make it available for the bioscience research community.

The workflows are generic and work well with glycans, drugs and enzyme substrates. However, although the developed workflow has broad potential, in the context of this paper, we will focus on carbohydrate (glycan) recognition: a phenomenon critical to a number of biological functions in humans including highly specific responses of the immune system [1 – 3] and interactions of sugars with biosynthetic (glycosyltransferase) and degrading (glycosidase) enzymes [4, 5]. Protein-glycan interactions are keys in cancer cell biology and in recognition of bacteria and viruses [6 – 8]. Indeed, perturbation of this binding has therapeutic potential. Unlike polypeptides and proteins, oligosaccharides are highly dynamic molecules that may occupy different conformations over time and space. Understanding their conformations should provide clues towards the mechanisms which lead to specific and selective recognition of carbohydrates by a range of proteins [9, 10]. Our current understanding of biological intermolecular responses is dominated by mechanisms involving cell surface proteins, however it is now well documented that complex carbohydrates (often associated with membrane proteins or lipids) are major contributors to the mechanism of specificity in biological recognition processes.

INTRODUCTION

The value of GRID computing

In silico tools that perform docking and molecular dynamics simulations, have significant potential to contribute to biomedical research. However, simulations are computationally intensive and even with relatively small and less complex molecules can take weeks to complete on a single PC. As the "jobs" are repetitive in nature and could be split into "sub-jobs", the simulations can be run in parallel and the most obvious way to speed up the computation is to use computer clusters or the GRID [11]. In our developed workflow, the jobs are ported to and run on either the University of Westminster cluster or the UK National Grid. The computer scientists in the team have provided seamless access to computer and data resources available on the UK National Grid Service (NGS) using a P-GRADE grid, a web based, service rich environment for the development, execution and monitoring of workflows and workflow based parameter studies on various grid platforms, to run these simulations. Where appropriate, we have also exploited the Grid portal to access additional resources scattered amongst various other Grids, such as the EU Enabling Grid for E-SciencE (EGEE), the US Open Science Grid (OSG) and the US TeraGrid (TG), which also support international co-operation between bioscientists.

In our workflow, we combined GROMACs (a molecular dynamics simulation package) [12] with AutoDock (http://w3.to/autodock) which uses a search algorithm (*e.g.* Genetic Algorithm, and Monte Carlo) in combination with a free-energy desolvation function to provide a quick method for identification of putative active/binding sites [13]. The availability of such programs, run on parallel systems with a web-interface, would enable bioscientists to focus their resources and better plan experiments by allowing them to visualise potential interactions and determine the best molecules to investigate in the wet laboratory. This would reduce time and cost and increase the numbers of molecules screened. The ProSim project incorporated both application development and application execution.

Application development

Computer scientists migrated the protein molecule simulation applications as legacy codes to the Grid creating Grid services from these applications [14]. They also developed workflow templates representing different low-level user scenarios and/or applications specific portlets corresponding to high-level user scenarios.

Application execution

Bio-scientists execute the workflow templates through the Grid portal. These workflows are built on workflow templates that allow bio-scientists to modify the templates, create their own workflows, parameterise and run them. Application portlets hide the workflow-level details and enable users to define their input parameters, run and monitor their experiments, and capture and visualise the results through applications portlets. The Grid portal offers both parameter study and computational workflow support. These are essential to these simulations, considering the wide value range of input parameters and the large number of simulation steps involved.

Molecular Dynamics (MD) simulation packages such as AMBER (http://ambermd.org/; [15]) and CHARMM (http://www.charmm.org/; [16]) can provide quantified data for time dependant conformational behaviour of dynamic biological molecules such as complex carbohydrates. However these programs run slowly and they are commercial software packages which raise licensing issues on the Grid. As a result in our prototype we used GROMACs [12] which is an open-source software. It gives acceptable speed and allows the use of force fields from other molecular dynamics (e. g. AMBER) packages. AutoDock (http://autodock.scripps.edu/) which uses a search algorithm in combination with an empirical binding free energy desolvation function provides a quick method for identification of putative active/binding sites in recognition. Molecular dynamics can then be used to dissect the mechanism of recognition after docking and hence provide insight into the physiochemical properties which provide biological phenomena with remarkable specificity in recognition processes.

Simulation packages which can provide insight into biological recognition processes have significant potential to contribute to biomedical research if the results of the simulation can prove consistent with the outcome of conventional wet laboratory experiments (Figure 1). The aim of the ProSim project was to optimise and develop workflows consisting of command line driven software applications which are executable through simplified web-based interfaces to serve the purpose of either predicting or informing/directing wet laboratory procedures for biologists with limited computing skills.

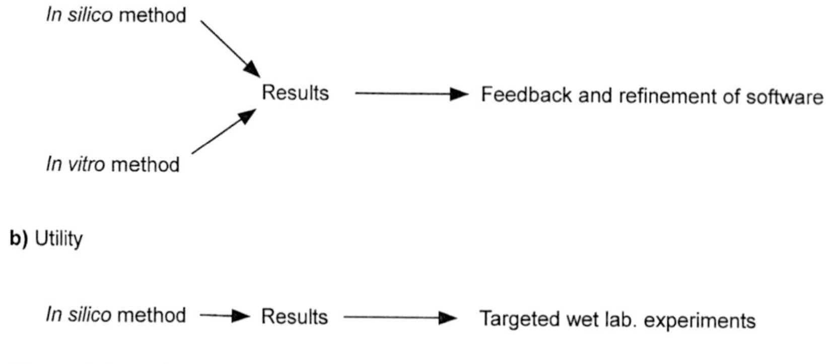

Figure 1. Experimental plan

ProSim Approach

To achieve this aim, we are validating the use of *in silico* modelling in determining how proteins interact with ligands and how to manipulate them to improve or change their specificity. To do this, we have integrated readily available software programs in an optimised workflow to reproduce receptor-ligand complexes with a good degree of accuracy (Figure 2).

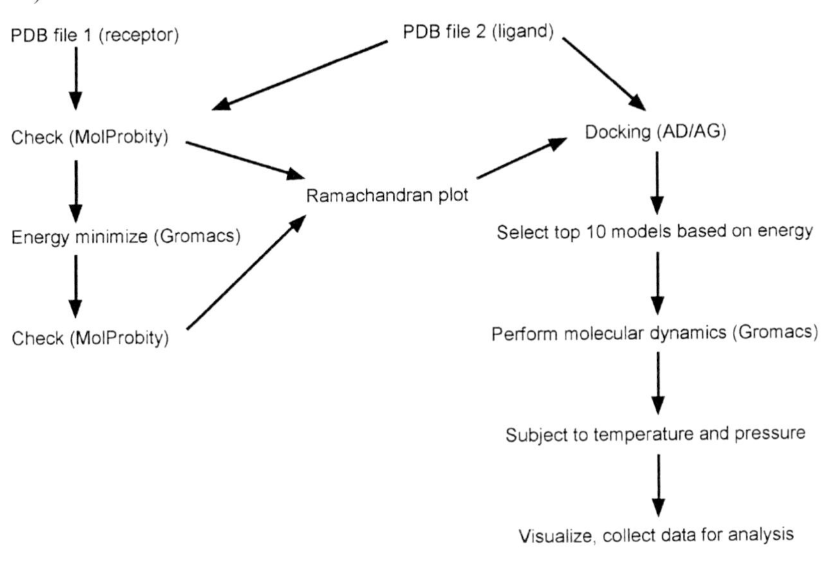

Figure 2. A schematic of the modelling process

In the workflow, the simulation process is divided into four phases: phase 1 involves selecting and preparing the receptor, phase 2 is selecting and preparing the ligand, docking the ligand to the receptor is performed in phase 3 and refining of the ligand-receptor molecule is carried out in phase 4. A schematic, showing how the output of one process is fed automatically into the next is shown in Figure 3. The boxes represent jobs which can take inputs through input ports and produce outputs through output ports (represented by the tiny boxes with the numbers). Application developers can connect programs through these ports to create workflows. The user does not have to deal with storage and program execution issues.

Phase 1 and 2: selecting and preparing receptor and phase ligand

Phase 1 and 2 incorporate the same operation: selection, pre-processing, solvation, energy minimisation and validation of the selected component in order to prepare the components required for docking. The pdb files contain Cartesian atomic coordinates which define 3-D models of organic and organo-metallic compounds (Bernstein *et al.* 1977). These Cartesian coordinates allow 3-D model visualisation using computer assisted software such as RasMol (http://rasmol.org/; [17]) and Swiss PDB viewer deep view (http://spdbv.vital-it.ch/).

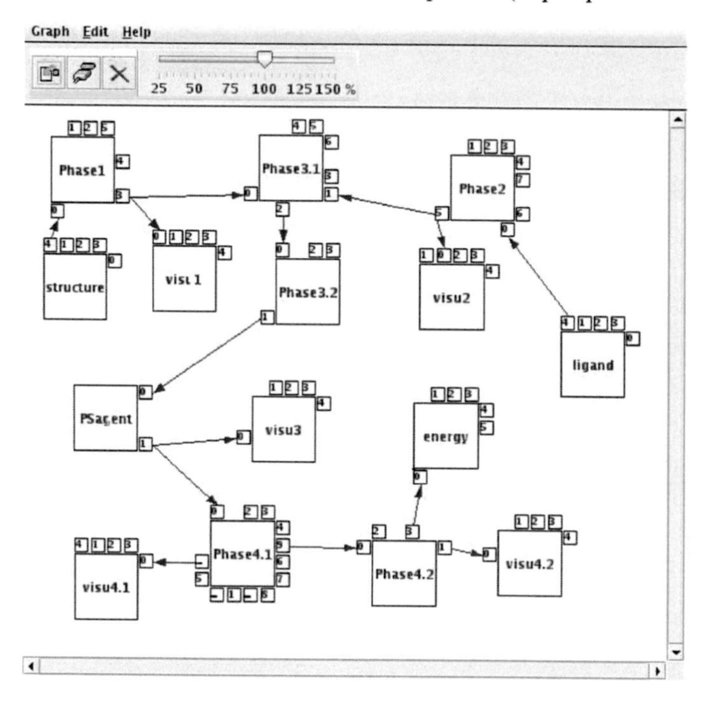

Figure 3. The phases of the workflow.

There are a number of different methods available which can be used to retrieve or generate pdb coordinates for various compounds. For complex biological compounds with thousands of atoms, the pdb coordinates are most frequently derived from data generated using X-ray crystallography and NMR spectroscopy. These pdb coordinates can be downloaded from a central database at http://www.rcsb.org. A second method for generating 3-D structures of complex biological compounds is through *in silico* homology modelling or fold recognition for which there are also a number of servers available such as SWISS MODEL [18] and PHYRE [19].

For simpler compounds, 3-D structures in pdb format can be generated using Smiles (http://www.molecular_networks.com/products/corina) coding strings. This method produces a rough structure which then requires further refinement or, through the use of biomolecular force fields, the introduction of appropriate parameters. The use of the appropriate force field allows optimisation of the potential energy and overall 3-D topology (such as partial atomic charges, bond length and torsions) of the compounds. Force field parameters can either be based on quantum chemistry calculations and/or defined by wet laboratory experimental observations (i. e. X-ray/NMR data).

The downloaded pdb files are not always correct and often require a number of pre-processing steps prior to use in experiments. This may include tasks such as removing coordinates for atoms which are not physiological components of the selected receptor but by-products of reactions used in solving the structure. The pdb coordinates probably do not represent the structure of the receptor at the energy minima and therefore require some form of optimisation. Although structures are often refined prior to deposition in the database, it is important to perform energy minimisation in order to relax the structure. This can be carried out in the presence of the required physiological solvent and ions. Although this step does not make the pdb model 'more accurate', it does correct any steric hindrance and repairs distorted geometries in the 3-D structure by releasing internal constraints. This step is particularly useful if the initial model has been manually manipulated (i.e. through the introduction of a mutated residue or addition of new bonds). In effect, the manipulated atoms are repositioned according to the defined force field used in the energy minimisation step and consequently produce a refined structure at the energy minimum. Changes in the position of atoms can be analysed by comparing the root mean squared deviation (RMSD) of the original with the energy minimised structure.

The structure is then checked using programs such as PROCHECK [20] or MolProbity [21]. For complex biological compounds such as proteins, a pdb file containing 3-D coordinates of the selected compound needs to be checked or validated. One way to assess the quality of the receptor structure (for proteins) is to produce a Ramachandran plot which describes the conformations of the individual building units of the protein molecule. The plot can give a good estimate of how "correct" the pdb coordinates are in comparison with that which is already known about protein structure [22].

Phase 3: docking the ligand to the receptor

The docking program used in the protein simulation is AutoDock. It performs the docking in two stages. First, it allows the ligand to move in geometrical space until contact with the receptor is made. Next, AutoDock employs a semi-empirical desolvation force field equation to calculate binding free energy. The lowest binding free energy is identified by a number of iterations (defined by the user) based on either genetic or simulated annealing algorithms. Finally, AutoDock ranks the docked conformations according to their lowest binding free energy. This method has been shown to predict and reproduce X-ray data with a good degree of accuracy [23].

The results generated after AutoDock can be either visualised or analysed using different tools. A visual 3-D model showing hydrogen bond formations between the receptor and the ligand can be produced. Since AutoDock utilises a scoring function, the data can also be clustered to show conformations of the ligand which interacts with the receptor through similar binding free energies.

Phase 4: refining the ligand-receptor molecule

AutoDock often produces a number of different ligand conformations when in contact with a receptor which are found within the same binding free energy. In order to map the atomic interactions between the ligand and the receptor, the molecular dynamics (MD), approach is used downstream of the docking experiment. Through performing MD on different docked models, as defined by their binding free energy data, the most stable complexes can be identified as defined by the MD parameters. This can then be tested in the laboratory for validation through biochemical approaches.

Benefits and Limitations of our Approach

In addition to developing such workflows, by parallelisation we can take advantage of significant reduction in computational time and generate more simulations which enable statistical analysis of larger data sets over shorter periods of time that should yield better models. An important outcome of our approach would be to enable biologists to identify limitations in computational chemistry software and consequently report back to developers to enable evolution and improvements in software packages in biological contexts.

Using the workflow

The biologist needs to log into the GRID system and then open the ProSim workflow. They can then navigate to a page where they upload their pdb files and set the interaction box size. It is important that the box is not too small as the molecules need to be completely encompassed by it, on the other hand if the box is too big, the molecules will not encounter each other during the experiment. The workflow is then started and the jobs are sent to nodes on the GRID for completion and return of the finished data to the researcher. Throughout the workflow, progress can be monitored and results of workflow jobs can be viewed; this enables jobs with errors to be "spotted" and halted or refined. The researcher then has all the data to create models, analyse interaction parameters and further process the data. For more experienced users, there is the facility to refine the workflow and introduce more specific parameters based on experience and previous data.

Validating the workflow

Analysis of the docking of a thiodisaccharide substrate analogue to a human family 47 glycosylhydrolase (mannosidase)

The structure of a human family 47 glycosylhydrolase (mannosidase) with a bound thio substrate analogue of an α-(1,2)-dimannan has previously been solved using X-ray crystallography (pdb file 1X9D). The thiodisaccharide is a substrate analogue that is cleaved slowly such that the binding of the substrate to the enzyme can be analysed. The natural substrate of the enzyme is not used as it is cleaved so quickly that it would not be seen in the crystal structure. As a proof of principle of our *in silico* methodology, a *de novo* constructed ligand structure was docked to the protein, using as target the crystal structure from which the analogue had been removed.

Preparation of the 1X9D pdb file

As there were some atoms missing in the coordinate file of the original crystal structure, in order to complete the file for submission the sequence was first submitted to the Swiss homology model server (http://swissmodel.expasy.org/; [18]). The molecule is shown as a ribbon representation in Figure 4. The molecule has a central calcium ion and, when viewed in 3D it is apparent that the catalytic site is situated at the bottom of a deep funnel-like channel. The terminal α-1 – 2 linked mannose is thought to be coordinated by the Ca^{2+} ion, bringing the linkage oxygen into the appropriate position for cleavage. Due to energy barriers on the way to the final position, the binding of the cleaved products is much weaker forcing the reaction products out of the catalytic centre of the enzyme.

Preparation of the thiodisaccharide coordinate file

The ligand file was prepared using a molecular editor and subsequent translation of the small molecule file into a pdb coordinate file as previously described. This output was fed directly into the PROSIM workflow.

Docking results for the disaccharide analogue and 1X9D

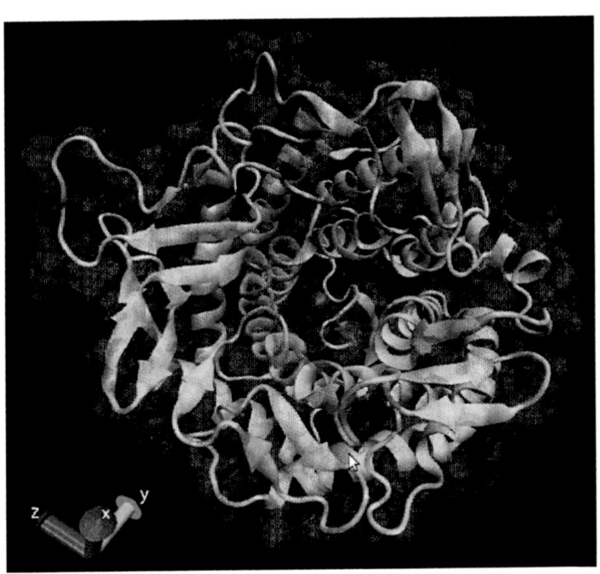

Figure 4. Ribbon representation of the human α-mannosidase 1X9D with the central calcium ion in vdW representation (red) and a transparent surface representation overlaid.

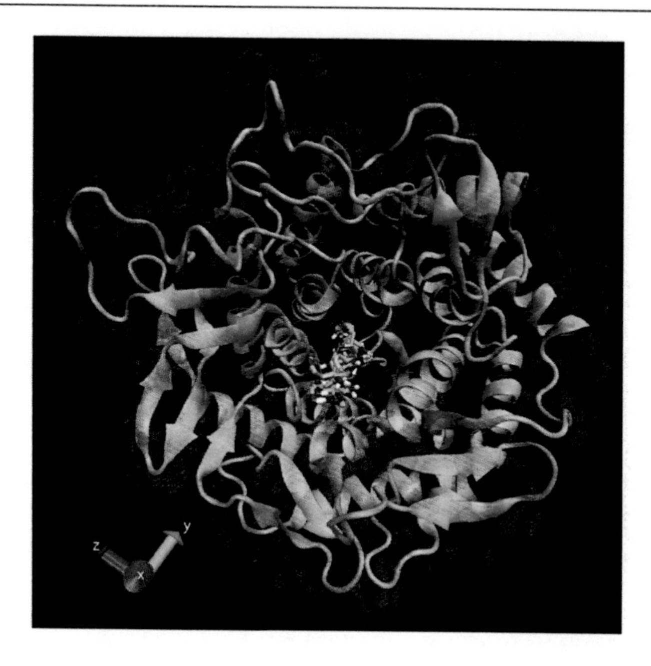

Figure 5. Five docking results shown together with the protein structure. The proper results are found in the third clustering group near the energy minimum.

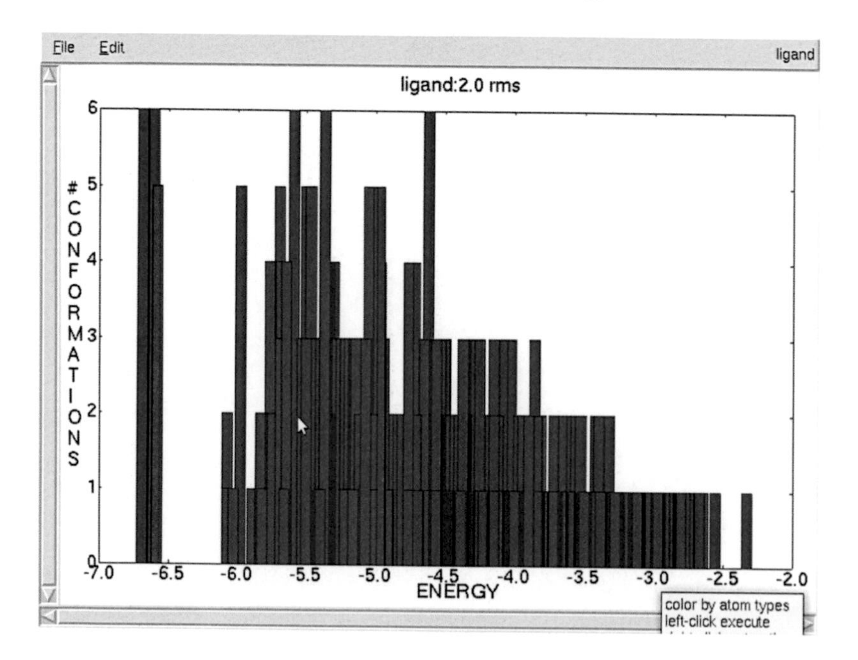

Figure 6. Clustering of the 1000 docking runs performed by the PROSIM workflow. The cluster containing the conformations in the proper position is marked red.

Figure 7. Superposition of the five Autodock results (in thin stick representation) and the thiodisaccharide from the original crystal structure.

The results of docking 5 analogues to the enzyme are shown in Figure 5. The automated workflow could not find the proper docking site using the blind dock option; twelve results showed an energetically more favourable result, but the difference was very small. It is apparent from the docking results (Figure 6) that due to lectin-like binding at alternative sites on the receptor surface further favourable docking sites may be found (especially in blind docking experiments). In blind docking wider grid spacing is used and hence the final positioning may not be as exact as if an approximate docking site was chosen on which docking could be focussed.

In Figure 7, the sugar ring at the position 1 in the catalytic centre (left) is very well aligned. The mannose at the position 2 (right) shows more variation. The position of the sulphur atom that is important for the catalytic mechanism is very well conserved. This shows that:

1. the *de novo* docking experiments reproduce the final position of the ligand found in the crystal structure
2. the central sulphur atom which marks the position of the linkage oxygen is well placed
3. there is much less conformational freedom for the central mannose residue than for the more distal one.

Comparison of the docking of oseltamivir to a H5N1 influenza virus neuraminidase and a Trichomonas vaginalis sialidase homology model

This trial aimed to demonstrate the potential binding of the antiviral drug oseltamivir (Tamiflu®) to the influenza neuraminidase and a homology model of a putative sialidase (neuraminidase) of the protozoan *Trichomonas vaginalis*. The binding of the drug to the influenza neuraminidase is well documented [24]. The aim was to validate the workflow using a known enzyme: inhibitor pair and to determine whether the protozoan enzyme bound the inhibitor and if so, what residues were responsible for that binding.

Preparation of the ligand pdb file

The ligand file was constructed by drawing a 2D sketch of the small molecule using JME a java based molecular editor (http://www.molinspiration.com/jme/index.html). The 2D construct was then translated into a 3D coordinate file using e. g. a SMILES translator or the program CORINA which is available on the molecular networks homepage (http://www.molecular-networks.com/products/corina). The ligand coordinate file generated by the SMILES translator was used as an input file for the ProSim workflow. A blind dock was performed to scan the receptor for possible low energy binding sites.

Preparation of the viral neuraminidase enzyme .pdb file

The protein coordinates were taken from the RCSB server (2HT7). As before missing residues and atoms were added by submission to a homology model server and the resulting pdb file submitted to the PROSIM workflow.

Preparation of the *T. vaginalis* enzyme pdb file

Using known eukaryotic sialidase sequences the *T. vaginalis* genome (www.trichdb.org) was searched for putative sialic acid cleaving enzymes. The applied domain recognition software (*e.g.* Pfam) suggested the existence of a six blade beta propeller fold which is typical for sialidases. In a next step the most promising sequences were fed into an automated homology model server and the returning results were critically reviewed with respect to structure probability and completeness. The arginine triad typical for sialidases was taken as a second “identifier” of the sialidase structure. Since this work was begun, the selected genes have been annotated out of the genome and classified as sialidases. The enzyme pdb file as it was generated by the Swiss homology model server was the input into the ProSim workflow.

Figure 8. Ribbon representation of the influenza viral (H5N1) neuraminidase 2HT7. The characteristic arginine triad is shown in red stick representation and the molecular surface in transparent representation.

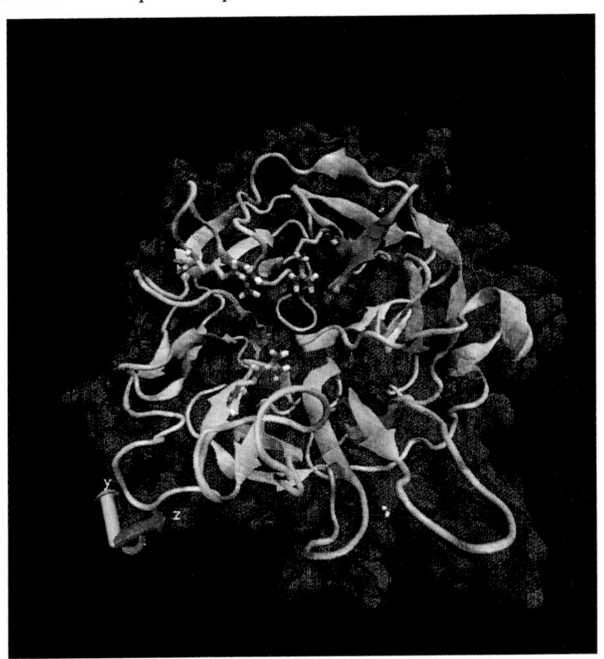

Figure 9. Visualization of the putative *T. vaginalis* sialidase showing a transparent molecular surface rendering the underlying secondary protein structure in cartoon mode and the arginine triad (blue) typical for this class of enzymes. The six-fold symmetry of the propeller structure is also seen.

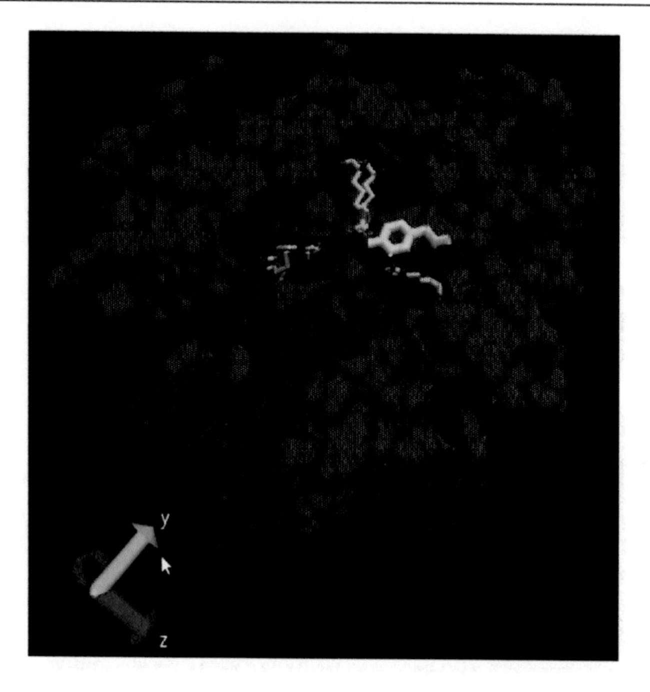

Figure 10. An alignment of the *T.vaginalis* sialidase and the H5N1 viral neuraminidase with the arginine triad in the proper position and the tyrosine residue (Y264) which blocks the access to the active centre and does not exist in the protozoan enzyme coloured in yellow.

Docking results for the viral Neuraminidase

The execution of the workflow typically takes less than 24 h. 1000 docking trials were performed and the 10 best conformations conserved. Of these ten conformations consecutive short molecular dynamic runs at room temperature 1 bar pressure, explicit solvent and ion concentration of 150 mM were performed.

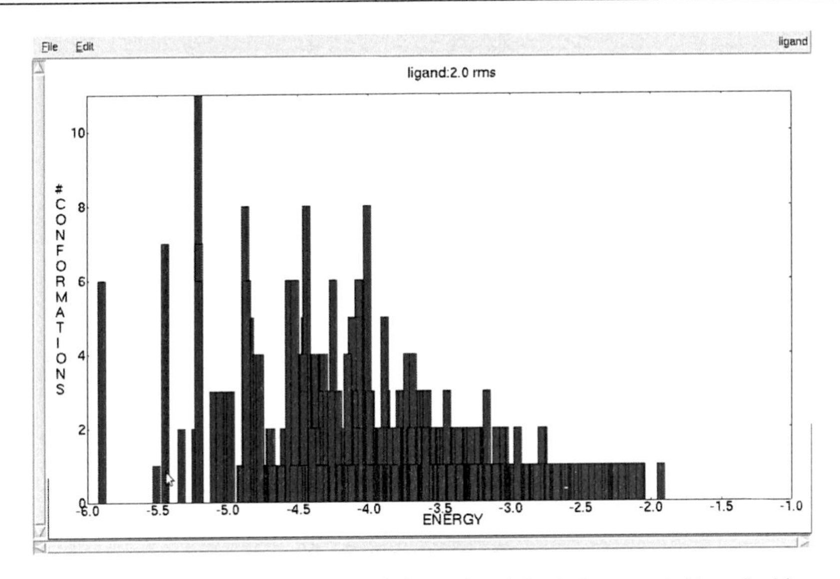

Figure 11. Clustering of the results of the oseltamivir viral neuraminidase docking. The red bar and the two neighbouring bars give reasonable results near the active centre of the enzyme.

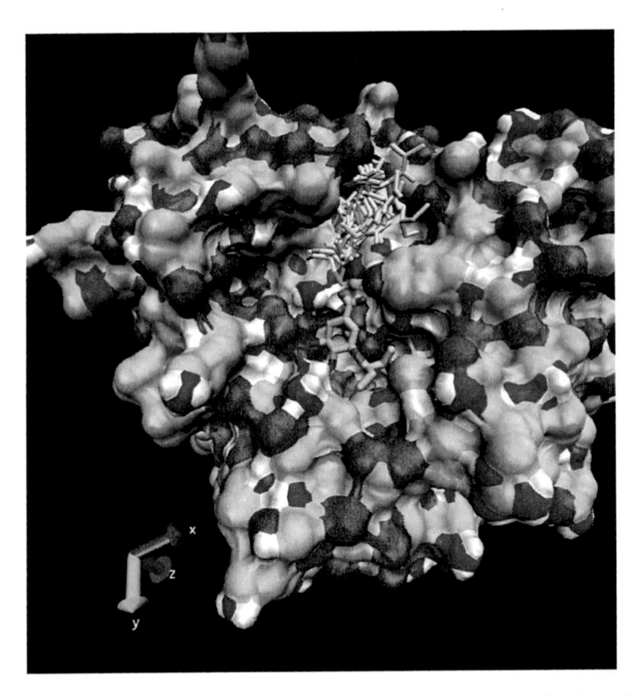

Figure 12. H5N1 viral neuraminidase with the substrate as localized in the crystal structure (stick representation near the centre of the molecule and with the ten docked results which are found near the active centre which is formed by the arginine triad.

The results of the blind docking of oseltamivir to the viral neuraminidase surprisingly do not agree with those of the crystal structure (see Figures 11 and 12). This is difficult to explain as it has been claimed that the antivirals oseltamivir and zanamivir have been designed using *in silico* methods. Considering the results of another group (Wang *et al.* 2010) the ligand seems to have docked in the region of the second substrate sugar whilst the active centre seems to be "unreachable" for the docking algorithm

A possible reason is a flexible tyrosine (Y264) that may, *in silico* impede the access to the active centre but may in reality be part of a flexible receptor system (induced fit). This residue is not found in the *T. vaginalis* enzyme. To prove this, the tyrosine in the viral neuraminidase was mutated to an alanine and the submission to the PROSIM workflow repeated. A run with the enzyme with no added inhibitor was conducted in order to study possible spontaneous movements and flexibility of the tyrosine residue

Docking results for the mutated receptor structure

As expected when the blocking tyrosine residue was replaced by the less bulky and apolar alanine some of the docking results with a low energy level were in the expected site (Figures 13, 14, 15 and 16).

Figure 13. Critical tyrosine residue which may impede the docking if the receptor molecule is viewed as completely rigid.

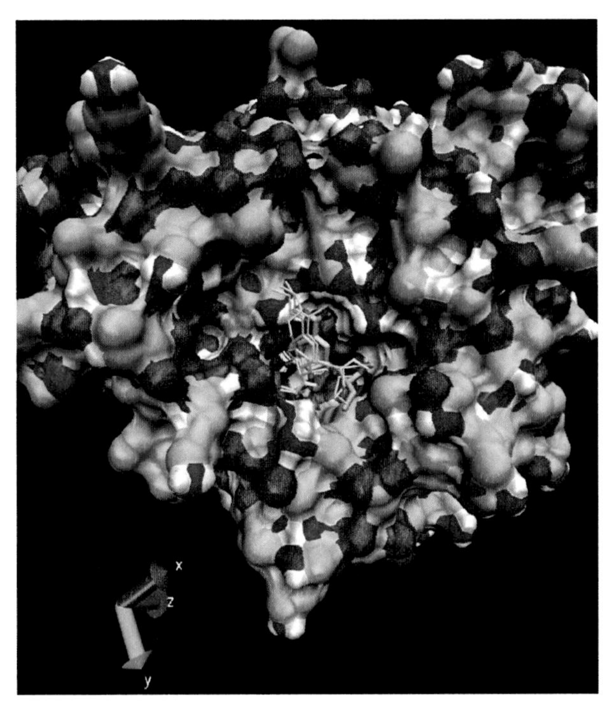

Figure 14. Four docking results placed properly by the Autodock algorithm (in yellow stick representation). The small molecule from the original crystal structure is shown in stick representation, the receptor in ribbon representation with a transparent surface.

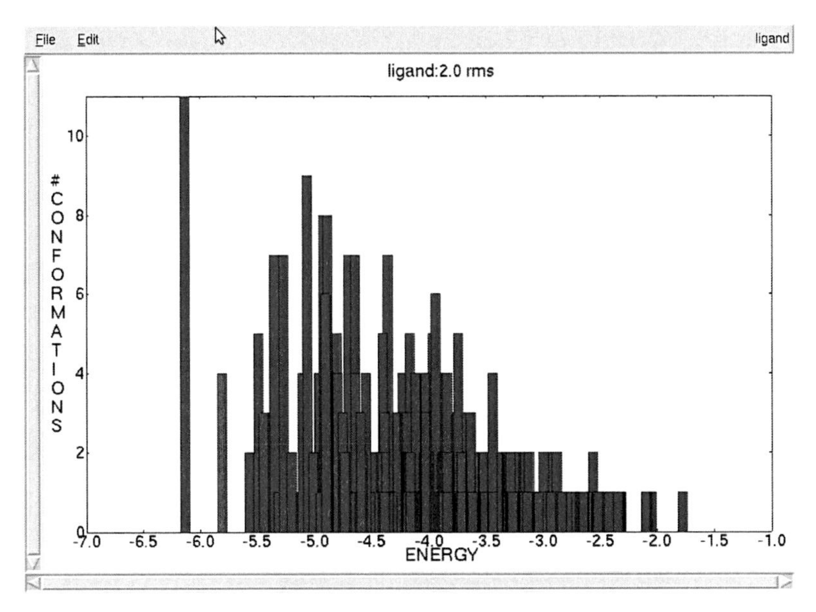

Figure 15. Clustering of the energy levels of the docking results for the mutated enzyme.

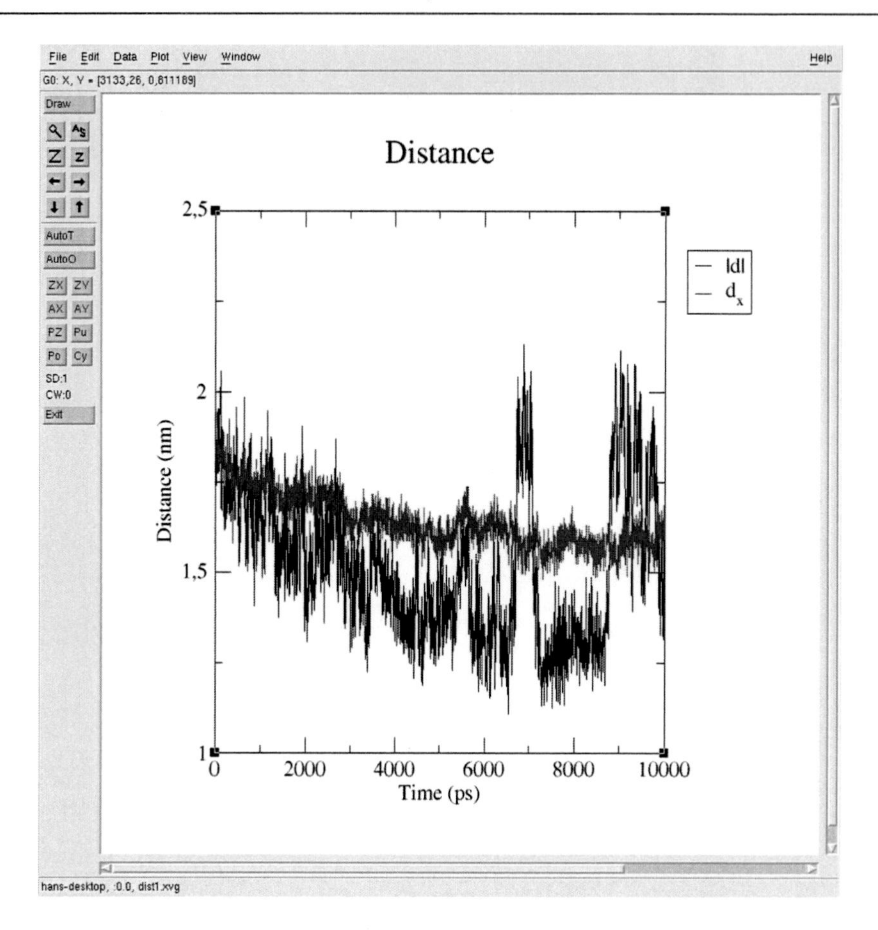

Figure 16. Distance between an atom near the centre of gravity of the receptor and the Y264 C-alpha (red) and the Y264 hydroxyl oxygen (black) showing the high mobility of the "lid".

The docking of oseltamivir to the protozoan sialidase

Sialidases of the sexually transmitted protozoon *T. vaginalis* facilitate the cleavage of terminal sialic acids which are frequently found in the mucus produced by epithelial cells and glands in the genitourinary tract of the host organism and are thought to play a role in cytolysis [25, 26]. The homology model of the sialidase was docked with the oseltamivir inhibitor and the results are shown in Figures 16 and 17. The ligand does not really enter the catalytic cavity but seems to block the putative half pipe like entrance that could impair substrate binding.

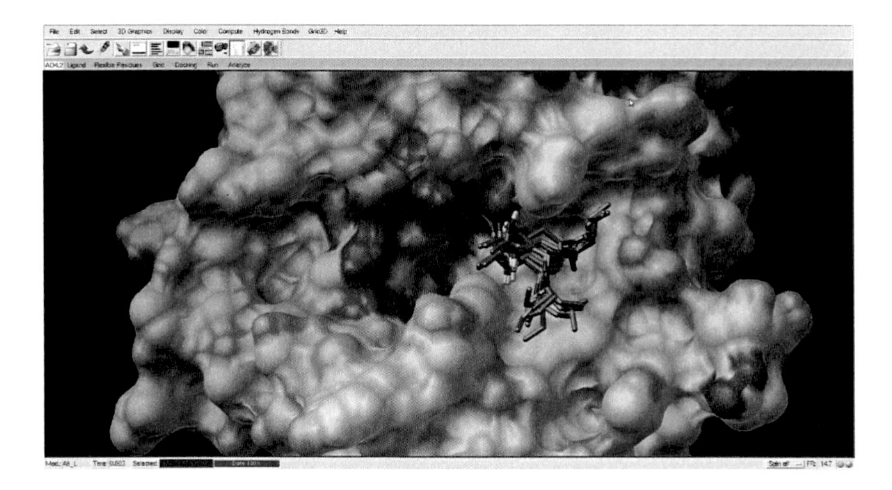

Figure 17. Six of the ten best docking results show the ligand in stick representation with the atoms coloured according to type and the receptor shown in surface representation. Note the arginine triad in blue left of the ligand.

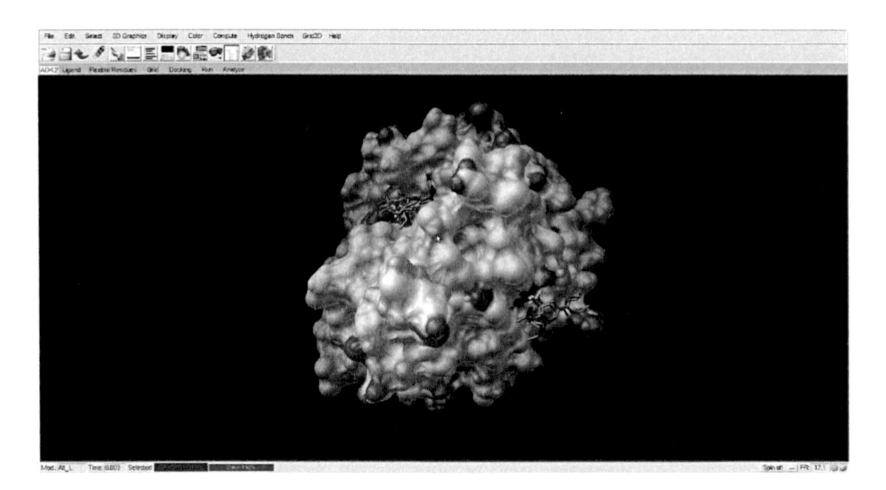

Figure 18. This figure shows a slightly turned version of Figure 17. The two ligand conformations visible in this picture represent the 10 best docking results from 1000 trials.

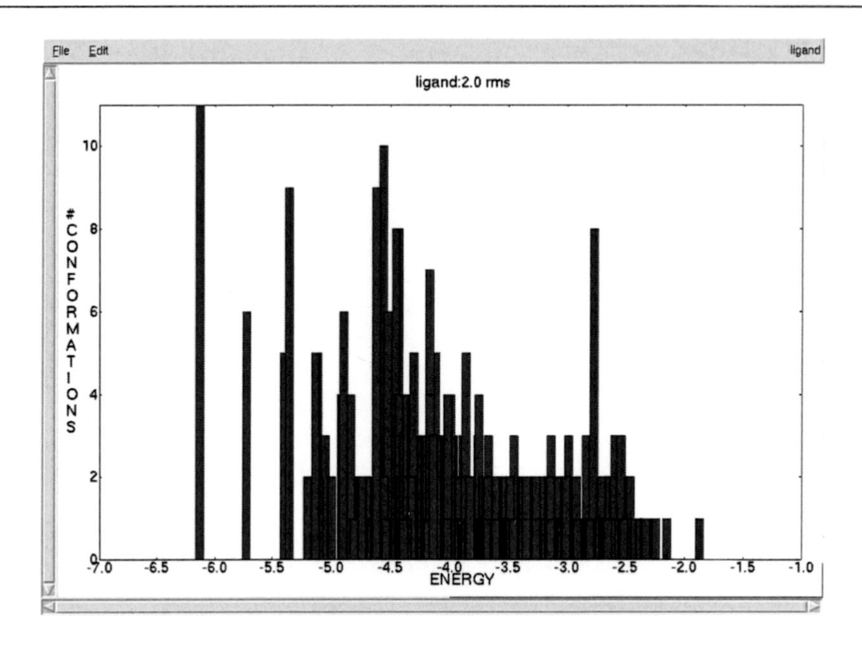

Figure 19. The histogram showing the "docking energy" (as named by the autodock convention) clustered in groups of max 2.0 rms. The group at the left represents the above shown conformations and can be well discerned from the bulk of results.

Analysis of the molecular dynamics trajectory

As a prerequisite the temperature and pressure profile of the molecular dynamics run is plotted to look for irregularities. For that purpose the GROMACs program *g_energy* was used and visualization performed using *xmgrace*.

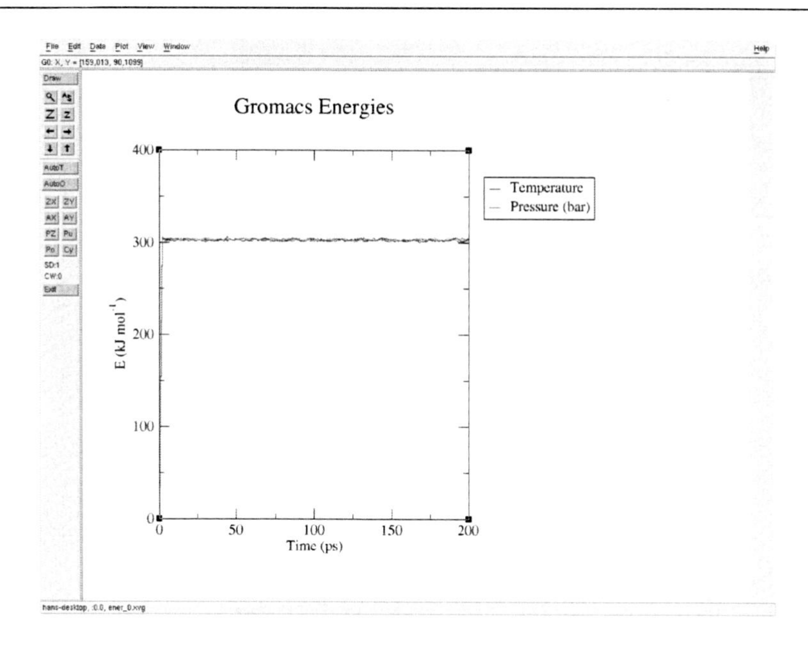

Figure 20. Temperature plot (overlay of 10 trajectories) showing constant temperature over the whole trajectory.

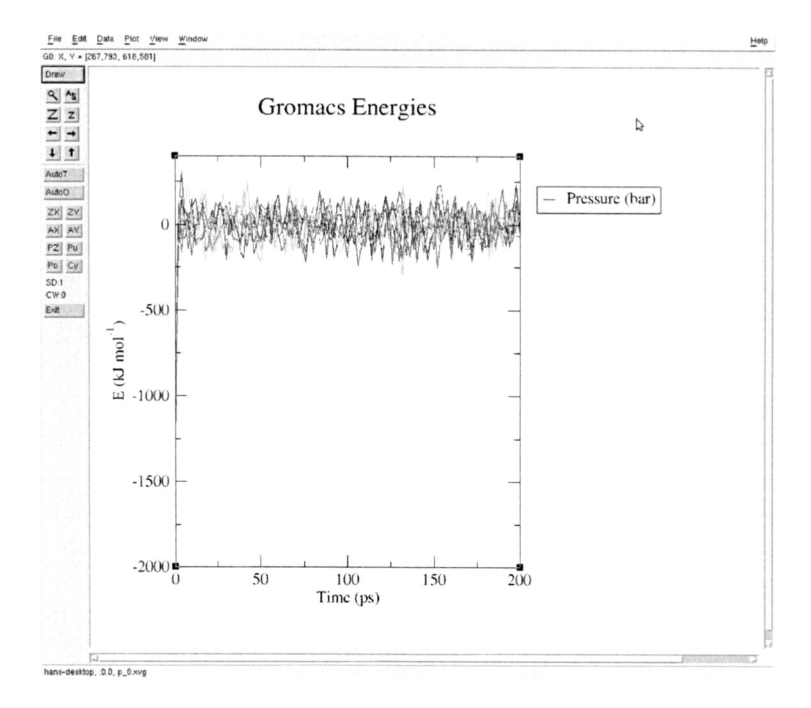

Figure 21. The pressure plot (overlay of all 10 trajectories) shows considerably more variation.

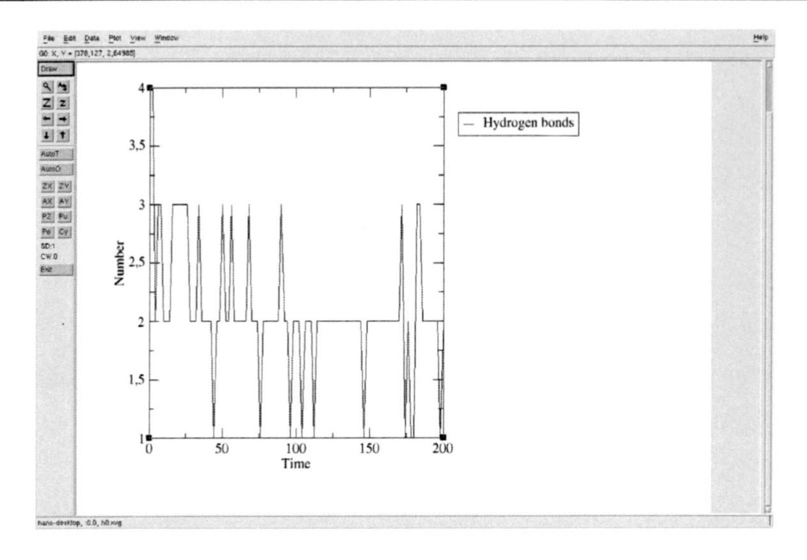

Figure 22. Number of H-bonds between the ligand and the enzyme as a function of time, generated with the function *g_hbonds* of GROMACs and visualized using *xmgrace*.

Conclusions

The PROSIM workflow is able to reproduce the docking of a sugar-like substrate very well as shown with the mannosidase. This example shows that sugars may also bind to other sites of the molecule with high affinity. A major improvement of the workflow would involve the user picking the most likely results and to exclude dockings with high affinity to areas other than the expected binding sites before starting the molecular dynamics run. An alternative would be to restrict the docking to regions of high interest. On the other hand the blind docking approach leads to more unbiased results. The second example involving the viral neuraminidase shows that AutoDock may fail if dynamic conformation changes take place during the binding or the access to the active site is blocked by a loop or a single residue. Interestingly although the *T. vaginalis* neuraminidase does not have an active site with a "lid", the docking into the active site fails, unlike the mutated viral neuraminidase. Here a molecular dynamics simulation exerting a force (push or pull) on the ligand could help to understand the conformational changes during the binding process. These results could also represent the differences between a eukaryotic and viral enzyme with respect to the artificial ligand oseltamivir which may be the basis of the low toxicity of the drug.

REFERENCES

[1] Zhang, X.L (2006) Roles of glycans and glycopeptides in immune system and immune-related diseases. *Curr. Med. Chem.* **13**(10):1141 – 7. doi: http://dx.doi.org/10.1016/j.ejmech.2009.10.040.

[2] Crocker, P.R., Paulson, J.C., Varki, A. (2007) Siglecs and their roles in the immune system. *Nat. Rev. Immunol.* **7**(4):255 – 66. doi: http://dx.doi.org/10.1038/nri2056.

[3] Erbacher, A., Gieseke, F., Handgretinger, R., Müller, I. (2009) Dendritic cells: functional aspects of glycosylation and lectins. *Hum. Immunol.* **70**(5):308 – 12. doi: http://dx.doi.org/10.1016/j.humimm.2009.02.005.

[4] Henrissat, B., Sulzenbacher, G., Bourne Y. (2008) Glycosyltransferases, glycoside hydrolases: surprise, surprise! *Curr. Opin. Struct. Biol.* **18**(5):527 – 33. doi: http://dx.doi.org/10.1016/j.sbi.2008.09.003.

[5] Hinou, H., Nishimura, S. (2009) Mechanism-based probing, characterization, and inhibitor design of glycosidases and glycosyltransferases. *Curr. Top. Med. Chem.* **9**(1):106 – 16. doi: http://dx.doi.org/10.2174/156802609787354298.

[6] An, H.J., Kronewitter, S.R., de Leoz, M.L., Lebrilla, C.B. (2009) Glycomics and disease markers. *Curr. Opin. Chem. Biol.* **13**(5 – 6): 601 – 7. doi: http://dx.doi.org/10.1016/j.cbpa.2009.08.015.

[7] Narimatsu, H., Sawaki, H., Kuno, A., Kaji, H., Ito, H., Ikehara, Y. (2010) A strategy for discovery of cancer glyco-biomarkers in serum using newly developed technologies for glycoproteomics. *FEBS J.* **277**(1):95 – 105. doi: http://dx.doi.org/10.1111/j.1742-4658.2009.07430.x.

[8] Reading, P.C., Tate, M.D., Pickett, D.L., Brooks, A.G. (2007) Glycosylation as a target for recognition of influenza viruses by the innate immune system. *Adv. Exp. Med. Biol.* **598**:279 – 92. doi: http://dx.doi.org/10.1007/978-0-387-71767-8.

[9] Shaikh, F.A., Withers, S.G., (2008) Teaching old enzymes new tricks: engineering and evolution of glycosidases and glycosyl transferases for improved glycoside synthesis. *Biochem. Cell Biol.* **86**(2):169 – 77. doi: http://dx.doi.org/10.1139/007-149.

[10] Woods, R.J. (1998) Computational carbohydrate chemistry: what theoretical methods can tell us. *Glycoconj. J.* **15**(3):209 – 16. doi: http://dx.doi.org/10.1023/A:1006984709892.

[11] Kacsuk, P., Sipos, G. (2005) Multi-Grid, multi-user workflows in the P-GRADE Grid Portal, *Journal of Grid Computing* **3**(3 – 4):221 – 238. doi: http://dx.doi.org/10.1016/j.future.2008.02.008.

[12] Van Der Spoel, D., Lindahl, E., Hess, B., Groenhof, G. Mark, A.E., Berendsen, H.J. (2005) GROMACS: fast, flexible, and free. *J. Comput. Chem.* **26**(16):1701 – 18. doi: http://dx.doi.org/10.1002/jcc.20291.

[13] Morris, G.M., Goodsell, D.S., Halliday, R.S. Huey, R. Hart, W.E., Belew, R.K. Olson, A.J. (1998) Automated docking using a Lamarckian genetic algorithm and empirical binding free energy function. *J. Computational Chemistry* **19**:1639 – 1662. doi: http://dx.doi.org/10.1002/(SICI)1096-987X(19981115)19:14<1639.

[14] Delaitre, T., Kiss, T., Goyeneche, A., Terstyanszky, G., Winter, S., Kacsuk, P. (2005) GEMLCA: Running legacy code applications as grid services. *Journal of Grid Computing* **3**:75 – 90. doi: http://dx.doi.org/10.1007/s10723-005-9002-8.

[15] Case, D.A., Cheatham, T.E., Darden, T., Gohlke, H., Luo, R., Merz, K.M., Onufriev, A., Simmerling, C., Wang, B., Woods, R.J. (2005) The Amber bio-molecular simulation programs. *Journal of Computational Chemistry* **26**:1668 – 1688. doi: http://dx.doi.org/10.1002/jcc.20290.

[16] MacKerell, A.D., Brooks, C., Brooks, L., Nilsson, L., Roux, B., Won, Y., Karplus, M. (1998) CHARMM: *The Energy Function and Its Parameterization with an Overview of the Program*. In *The Encyclopedia of Computational Chemistry* **1**:271 – 277. Schleyer, P.v.R. *et al.*, editors. John Wiley & Sons: Chichester. ISBN: 0 – 471 – 96588-X

[17] Bernstein, F.C., Koetzle, T.F., Williams, G.J., Meyer, E.F., Brice, M.D., Rodgers, J.R., Kennard, O., Shimanouchi, T., Tasumi, M. (1977) The Protein Data Bank. A computer-based archival file for macromolecular structures. *Eur. J. Biochem.* **112**:535 – 42. doi: http://dx.doi.org/10.1016/S0022-2836(77)80200-3.

[18] Arnold, K., Bordoli, L., Kopp, J., Schwede, T. (2006) The SWISS-MODEL workspace: a web-based environment for protein structure homology modelling. *Bioinformatics* **22**(2):195 – 201. doi: http://dx.doi.org/10.1093/bioinformatics/bti770.

[19] Bennett-Lovsey, R.M., Herbert, A.D., Sternberg, M.J., Kelley, L.A. (2008) Exploring the extremes of sequence/structure space with ensemble fold recognition in the program Phyre. *Proteins* **70**(3):611 – 25. doi: http://dx.doi.org/10.1002/prot.21688.

[20] Morris, A.L., MacArthur, M.W., Hutchinson, E.G., Thornton, J.M. (1992) Stereochemical quality of protein structure coordinates. *Proteins* **12**(4):345 – 64.
doi: http://dx.doi.org/10.1002/prot.340120407.

[21] Lovell, S.C., Davis, I.W., Arendall, W.B., de Bakker, P.I., Word, J.M., Prisant, M.G., Richardson, J.S., Richardson, D.C. (2003) Structure validation by Calpha geometry: phi, psi and Cbeta deviation. *Proteins* **50**(3):437 – 50.
doi: http://dx.doi.org/10.1002/prot.10286.

[22] Ramachandran, G.N., Ramakrishnan, C., Sasisekharan, V. (1963): Stereochemistry of polypeptide chain configurations. *Journal of Molecular Biology* **7**:95–9.
doi: http://dx.doi.org/10.1016/S0022-2836(63)80023-6.

[23] Englebienne, P., Fiaux, H., Kuntz, D.A., Corbeil, C.R., Gerber-Lemaire, S., Rose, D.R., Moitessier, N. (2007) Evaluation of docking programs for predicting binding of Golgi alpha-mannosidase II inhibitors: a comparison with crystallography. *Proteins* **69**(1):160 – 76.
doi: http://dx.doi.org/10.1002/prot.21479.

[24] Collins, P.J., Haire, L.F., Lin, Y.P., Liu, J., Russell, R.J., Walker, A., Skehel, J.J., Martin, S.R., Hay, A.J., Gamblin, S.J. (2008) Crystal structures of oseltamivir-resistant influenza virus neuraminidase mutants. *Nature* **453**:1258 – 1261.
doi: http://dx.doi.org/10.1038/nature06956.

[25] Padilla-Vaca, F., Anaya-Velázquez, F. (1997) Biochemical properties of a neuraminidase of *Trichomonas vaginalis*. *J. Parasitol.* **83**(6):1001 – 6.
doi: http://dx.doi.org/10.2307/3284352.

[26] Wiggins, R., Hicks, S.J., Soothill, P.W., Millar, M.R., Corfield, A.P. (2001) Mucinases and sialidases: their role in the pathogenesis of sexually transmitted infections in the female genital tract. *Sex. Transm. Infect.* **77**(6):402 – 8.
doi: http://dx.doi.org/10.1136/sti.77.6.402.

Glyco-Bioinformatics – *Bits 'n' Bytes of Sugars*
October 4th – 8th, 2009, Potsdam, Germany

GlycoOptimization for Fully Human and Largely Improved Biopharmaceutical Antibodies and Proteins

Steffen Goletz*, Antje Danielczyk, Renate Stahn, Uwe Karsten, Lars Stoeckl, Anja Loeffler, Annett Hillemann and Hans Baumeister

Glycotope GmbH, Robert-Roessle-Str. 10, 13125 Berlin, Germany

E-Mail: *steffen.goletz@glycotope.com

Received: 28th May 2010 / Published: 10th December 2010

Abstract

The vast majority of therapeutic proteins are, by nature, glycosylated. In the last years the attached glycans attracts more and more attention since an increasing number of data are available showing that glycosylation greatly affects the biopharmaceutical characteristics of the product. This article focuses on the novel GlycoExpress technology that allows to screen for the optimal glycosylation of any protein and to produce the glycooptimized biopharmaceutical for clinical use. GlycoExpress is a toolbox of human, glyco-engineered cell lines to express any protein with fully human glycosylation and in a variety of differentially glycosylated isoforms. In a number of *in vitro* and *in vivo* assays the optimally glycosylated protein is identified and the corresponding cell line is further developed for GMP-production of that glycooptimized biopharmaceutical.

Twelve proteins have been successfully glycooptimized so far with four antibodies and one protein hormone in late preclinical/early clinical development. Glycooptimization improved these biologics up to several hundred folds in respect to activity, bioavailability, immunogenicity and/or patient coverage. And it is expected that in the near future a number of novel biotherapeutics will be developed whose therapeutic or economic benefit rests upon a fully human and optimized glycosylation.

Introduction

Therapeutic proteins, such as antibodies, growth factors, protein hormones, cytokines, therapeutic enzymes, thrombolytic and blood coagulation factors, are produced by biotechnological techniques in genetically modified organisms (GMO) and assure the pharmaceutical and biotech industry excellent growth rates since many years. The vast majority of these therapeutic proteins are by nature glycosylated meaning that a number of different monosaccharides are attached to the amino acid sequence of the protein to form mainly species specific N-linked and O-linked glycans which contribute about 40% to the protein mass of for example the well-known erythropoietin (EPO) with tremendous impact on the potency of a protein.

The potential to improve these biopharmaceuticals by targeting its glycosylation is large. Therapeutically important characteristics of a protein, *e.g.* its activity, serum half-life, antigenicity and immunogenicity, its stability, solubility and even the productivity in a heterologous cell line, can be affected by the presence and/or the nature of the attached glycans [1, 2]. There is a wide range of mechanisms how the attached glycans contribute to the characteristics of a biopharmaceutical. To give some examples, glycans are synthesized by a portfolio of often species specific enzymes which results in a species specific glycosylation of the recombinant protein. In consequence biopharmaceuticals glycosylated in a non-human way bear the risk of immunogenic reactions as observed in several cases and in suboptimal performance in humans (for more details see below). The complex nature of hundreds of enzymes and transporters provided to assemble the glycan in a sequence of enzymatic reactions taking place after translation of the protein in certain compartments of the cell (ER, Golgi) explains why the productivity of a protein can be affected by its glycosylation. Negative charges introduced by the terminal monosaccharide sialic acid but also neutral monosaccharides influence intra- and intermolecular interactions, with the consequence that a protein is more or less stabilized or soluble or the binding characteristics of a ligand to its receptor or an enzyme to its substrate are modified. Glycans with terminal monosaccharide galactose and mannose-rich glycans are recognized for example by specific receptors in the liver that mediate a high clearance rate from circulation in the body. Certain therapeutic antibodies are much more active when glycosylated optimally and even the number of patients that is successfully treated (patient coverage) with that antibody depends on the type of glycosylation attached to the antibody due to a genetic receptor polymorphism in the population (for more details see below).

Since 2001 Glycotope GmbH develops and establishes novel technologies and products in the field of glycomics to improve biopharmaceutical proteins and combines today all the expertise to develop glycosylated biotherapeutics from the gene and the GMP production of the drug to clinical development in phase I and II trials. One focus was always to provide novel cell lines for production of improved biopharmaceuticals that are of human origin and allow the optimization of glycosylation.

Development of the GlycoExpress Technology for Glycooptimization of Therapeutic Proteins

There are several approaches to modify and optimize glycosylation:

1. The cell culture and production conditions are controlled to achieve a more reproducible and homogenous glycosylation.
2. The glycosylation is modified after production *in vitro* enzymatically.
3. The protein is glyco-engineered on the genetic level to add or eliminate sites for attachment of glycans.
4. A novel cell line is glyco-engineered for production of biotherapeutics with a modified glycosylation profile.

To optimize the glycosylation of a given protein at Glycotope all these approaches, except *in vitro* enzymatic modifications which are less suitable for pharmaceutical production, are used and if necessary combined to achieve a successful glycooptimization. However, in this article we will focus on the novel toolbox of human glyco-engineered cell lines (GlycoExpress), the successful glycooptimization using the GlycoExpress toolbox and the production of glycooptimized biopharmaceutical proteins in GlycoExpress cell lines.

Fully Human Glycosylation

One important aspect of glycooptimization is to achieve a fully human glycosylation of biopharmaceuticals [3]. Currently, human biopharmaceuticals are produced predominantly in *E. coli*, yeast or cell lines derived from insects (SF9), mice (SP2/0) or hamster (CHO). While a protein produced in different systems of bacterial or mammalian origin has at least the same primary structure, all post-translational modifications (PTMs) of this protein, most importantly the glycosylation, differ from organism to organism and even from cell type to cell type within one organism [4]. The mouse- and hamster-derived cell lines (such as SP2/0, CHO or BHK) used in industry and academia for production of glycosylated therapeutics are able to confer a glycosylation that has some similarity to a human glycosylation. However, important components found with human cells are missing (*e.g.* the 2,6-linked sialylation and the bisecting *N*-acetylglucosamine (GlcNAc)) and a number of non-human components have been found to significantly increase the likelihood of immunogenic reactions, such as terminal sialic acids that do not exist in human cells (*e.g.* NeuGc [5]) or terminal galactose linked to another galactose in a way that is absent from human cells (Gallili-Epitope). The latter has been found to be the major reason for the clinically observed severe hypersensitivity reactions in 33% of 72 patients in southern states of the USA treated with Erbitux® [6].

For this reason, Glycotope based its GlycoExpress technology on human cell lines providing human biopharmaceuticals with human PTMs and especially a fully human glycosylation. However, as mentioned above the glycosylation profile varies from cell type to cell type within one organism. It's therefore not surprising that a glycoprotein produced in different human cell lines is not glycosylated in the same way and that identifying the optimal glycosylation can result in proteins with largely improved therapeutic potency as compared to others. To screen for and produce proteins that are glycosylated in the optimal way, Glycotope glyco-engineered its human cell lines to achieve a set of cell lines with different glycosylation profiles.

Glyco-engineering of Cell lines to Achieve an Optimal Glycosylation

Since the glycosylation is a very complex process within the cell, the first step in glyco-engineering needs to be to understand and characterize the cell line specific glycosylation machinery. Therefore in a research program a large number of human cell lines was analysed by profiling them for the presence of important carbohydrate structures on the cell surface and for the presence of key enzymes of important glycosylation pathways on the mRNA and enzymatic level. Beyond this glycoprofiling it was important that a cell line chosen for glyco-engineering had excellent biotechnological features to allow the glyco-engineered cells to be used for production of the glycooptimized biotherapeutics.

For glyco-engineering of a chosen cell line several techniques are available:

1. Random and spontaneous or induced mutagenesis with phenotypic selection;
2. Genetic knock-out *via* site-specific recombination;
3. Stable transfection of glycosylation enzymes.

Figure 1 gives an example of a glyco-engineered cell line that express stably a novel glycan that was absent from the original cell line. In that case the relatively fast approach of induction of mutagenesis was successful to generate a small number of cells with the desired glycoprofile (see Figure 1, panel A) which could be selected, enriched and single-cell-cloned to establish the novel glyco-engineered cell line with the new phenotype (Figure 1, panel B). This glyco-engineering approach by induction of mutagenesis and phenotypic selection convinces by speed, good success rates and the lack of genetically modified organisms that are generated when the cells are genetically engineered as in approach 2 and 3. However, the phenotypic selection makes specific tools necessary that are not available for all glycostructures of interest. Therefore, the other two techniques are necessary add-ons in the repertoire of glyco-engineering techniques.

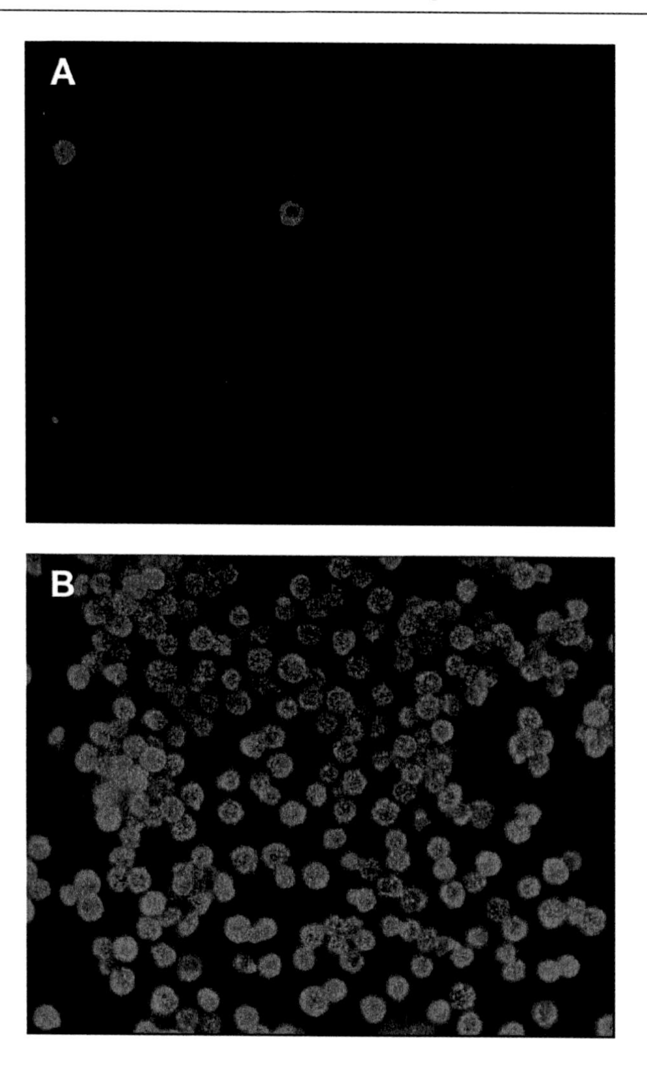

Figure 1. Cells stained for a desialylated key glycan analysed before (A) and after (B) glyco-engineering. In both panels the same cell number is shown.

A Toolbox of Glyco-engineered Human Cell Lines

The current set of cell lines that have been generated by glyco-engineering address the fucosylation, the α-2,3- and α-2,6-sialylation, the galactosylation, the antennarity and the presence of bisecting GlcNAc on glycoproteins. A given glycoprotein expressed in these cell lines can be produced *e.g.* in absence or presence of fucose, in absence or presence of 2,3- and 2,6-sialylation, at a very high degree of galactosylation (important for CDC activity of antibodies and to achieve high degrees of sialylation), in presence of bisecting GlcNAc and with antennarities from lower to higher complexity. To control the levels of sialylation, one cell line has been engineered, that is characterized by the capability to sialylate glycoproteins at a very high degree, an important feature to elongate the circulating half-life of a therapeutic glycoprotein. In addition, two glyco-engineered cell lines allow controlling the degree of sialylation or fucosylation to levels in between 0% and the naturally possible maximum by means of medium supplementation (metabolic engineering) and in-process control.

Hence, a glycoprotein expressed in these glyco-engineered cell lines can be made available in various glycoforms and screening in suitable human bioassays allows the identification of the particular glycosylation pattern that confers optimal product characteristics to the product. The whole procedure of glycooptimization is depicted in Figure 2 which also outlines another highlight of GlycoExpress to directly use one of the chosen GlycoExpress cell lines for development of a high yield production cell line for the GMP production process of the glycooptimized protein. The reason for the integration of screening and production are the excellent biotechnological features of the GlycoExpress cell lines which will be described in more detail below. In the following, two of twelve biotherapeutic products that were successfully glycooptimized are described in more detail.

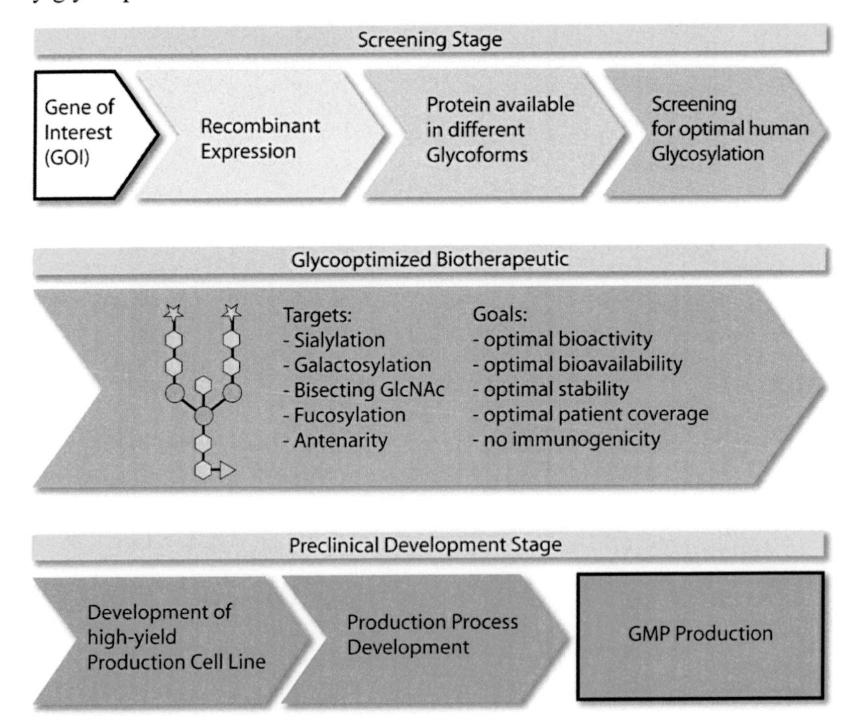

Figure 2. Schematic depiction of the integrated process of screening for glycooptimization and production of the glycooptimized product.

Glycooptimization of Antibodies

Antibodies are an important target for glycooptimization because of the remarkable economic success therapeutic antibodies have enjoyed in the last years and the observations that glycosylation can greatly affect the activity of an antibody *in vitro* and *in vivo*. Antibodies are glycoproteins with two N-linked glycans attached to the Fc part of IgG molecules plus additional glycans in the Fab protein in some of the antibodies.

Figure 3 A-C gives an example of an antibody that mediates tumour cell killing by activating natural killer (NK) cells and their antibody dependent cell cytotoxicity (ADCC) activity.

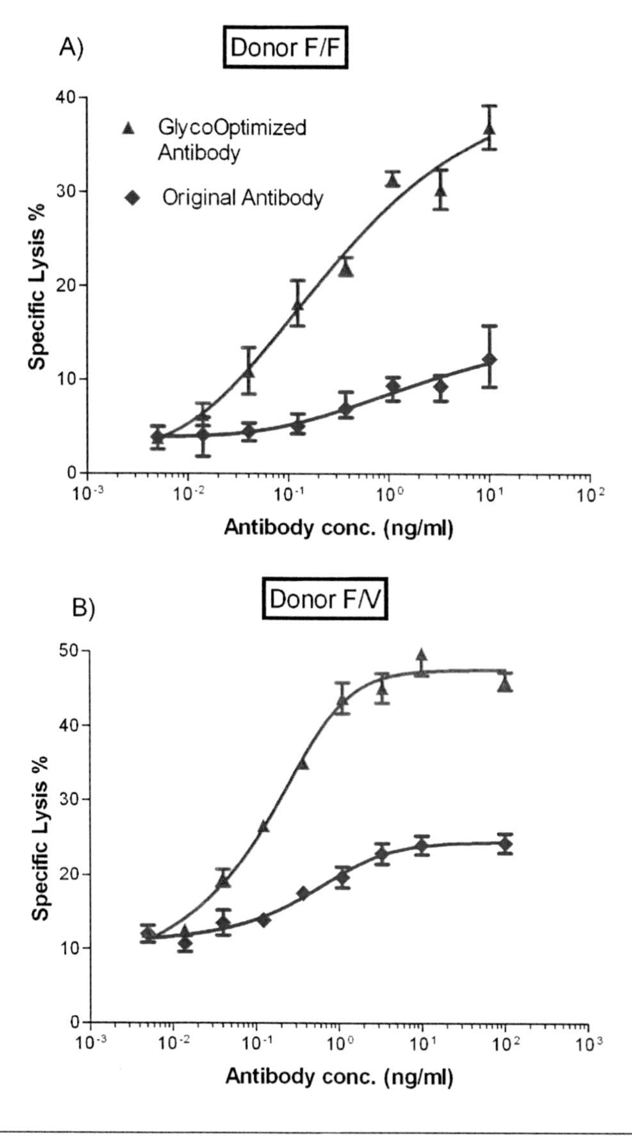

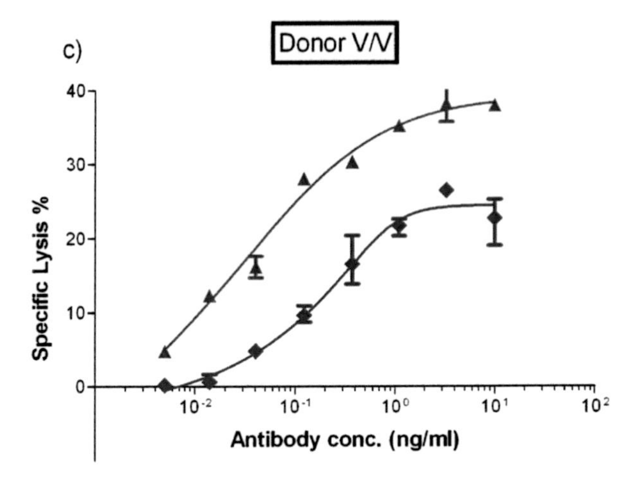

Figure 3. Glycooptimization of an antibody. The antibody-dependent cellular cytotoxicity (ADCC) is analysed *in vitro* for different concentrations of the original antibody (in blue) and the glycooptimized antibody (in red) using the cells of three donors of different FcγIIIa receptor phenotype F/F (A), F/V (B) and V/V (C).

By producing this antibody in the MabExpress toolbox of GlycoExpress cell lines, a fully human glycosylated and glycooptimized IgG1 antibody was selected with an up to 10fold to 250fold higher anti-tumour cytotoxicity (ADCC) compared to the original antibody produced in a rodent cell lines and depending on the donor type. Hence up to 250 fold less material is needed for achieving the same activity in human *in vitro* ADCC assays using human peripheral blood mononuclear cells (PBMC) as source for NK cells and human tumour cells as target. This is combined with an up to 4 fold increase in maximal tumour cell lysis activity, resulting in a potent ADCC anti tumour activity for all donor types with the three different FcγRIIIa ADCC receptor allotypes F/F, V/F and V/V (Fig. 3A-C). In consequence, the patient coverage is expected to be broadened by making the key anti-tumour activity (ADCC) available for all patients compared to about 20% when antibodies are produced in rodent systems (Table 1 and more details see below). In addition to the increase in potency, the half-life of the glycooptimized antibody was strongly elongated as shown in cynomolgous monkeys (data not shown). And the immunogenic carbohydrate moieties, such as the non-human Gallili-Epitope and non-human sialic acids present on the antibody from rodent production cells, which were reported to be the major reason for clinically severe hypersensitivity in up to 33% of the patients [6], was removed. These effects were obtained by combination of fucose removal and bisecting GlcNAc addition as well as a high sialylation and galactosylation degree and fully human glycosylation of the antibody product.

The ADCC increase by glycooptimization of the antibody is mainly caused by a much better binding and activation of the FcγRIIIa receptor by defucosylation of the antibody [7]. The FcγRIIIa receptor is located on NK cells and responsible for mediating the ADCC activity against tumour cells. A polymorphism of this receptor in the population causes the high variability of ADCC activity deviation of up to 10 – 250fold observed with the glycooptimized antibody. Figure 3 shows representative ADCC data for a glycooptimized antibody with PBMC from the various FcγIIIa allotypes and in comparison to the same antibody produced in rodent cells (SP2/0). The F/F and F/V phenotypes respond to the glycooptimized antibody 200fold/100fold better than to the original antibody (Fig. 3A and B), while donor cells of the V/V phenotype are less responsive to improvements by glycooptimization (10fold) still resulting in a clear increase in anti-tumour activity (Fig. 3C). This large increase of bioactivity especially with cells of F/F and V/F donors is largely explained by the fact that the antibody expressed in rodent cells is hardly able to activate the receptor of these cells because the binding affinity to the F allotype of the FcγIIIa receptor is much lower for the non-glycooptimized antibody. The responsible glycooptimization for the ADCC improvement is mainly due to a removal of 1,6-fucosylation and/or the addition of bisecting GlcNAc, both achievable with the GlycoExpress system. The original antibody produced in rodents is most active with PBMC of the V/V phenotype which correlates with a much better clinical outcome for V/V tumour patients in comparison to F/F and V/F patients for several non-glycooptimized antibodies such as Trastuzumab (Herceptin®), Rituximab (Rituxan®) and Cetuximab (Erbitux™) [8 – 10]. Glycooptimization increases the anti-tumour efficacy and makes it available not only for V/V patients who represent only 4 – 25 percent of the patient population but to all patients including the F/F and F/V phenotypes (Table 1).

Table 1. The distribution of the three different FcγIIIa receptor allotypes V/V, V/F and FF in the European and Asian population and its response to a glycooptimized antibody.

Receptor Type	V/V	V/F	F/F
Europe (in %)	8 – 27	48 – 60	33 – 49
Asia (in %)	4 – 11	35 – 47	43 – 54
ADCC increase (fold)*	10 x	100 x	250 x

* Compare with Figure 3A-C

This effect of strongly increased ADCC activity was observed with a number of antibodies when expressed in the MabExpress cell line. One of the antibody was largely ADCC improved not by removal of fucose but by addition of bisecting GlcNAc in combination with high degree of human α-2,3- and α-2,6-sialylation and galactosylation, showing that in contrast to the literature removal of 1,6-fucosylation is not the only reason for improvement of ADCC.

It will be very interesting to see the potency of these glycooptimized antibodies in the clinic. Currently among four glycooptimized antibodies in late stage development the first glycooptimized antibody, GT-Mab2.5-GEX, a novel, highly potent and tumour specific antibody recognizing 90% – 100% of all patients in all major carcinoma indications and multiple myeloma being effective already at 0.5 mg/kg in mouse tumour models, is in clinical trails.

Two other antibodies, GT-Mab5.2-GEX and GT-Mab7.3-GEX are glycooptimized and highly improved antibodies of which the originals are already marketed block busters (biobetters or largely improved 2nd generation antibodies). The clinical test for GT-Mab5.2-GEX has been started in the middle of 2010 and for GT-Mab7.3-GEX is expected to start early in 2011.

Glycooptimization of Non-antibody Glycoproteins

Not only antibodies can be successfully glycooptimized, also therapeutic proteins such as growth factors, glycoprotein hormones, cytokines, certain enzymes, blood factors and thrombolytica are all by nature glycosylated and good candidates for glycooptimization.

For example, a marketed growth factor currently produced in *E. coli* and yeast with either no glycosylation at all or irrelevant glycosylation lacking human sialylation was glycooptimized with optimized and fully human carbohydrates by GlycoExpress. Using the SialoFlex cell line various glycoforms with different degrees of human α-2,3- and α-2,6-sialylation where generated using metabolic engineering. SialoFlex has a defect at the key enzyme of the enzymatic pathway responsible for synthesis of human sialic acids. Therefore in a serum free production environment SialoFlex cells are lacking CMP-sialic acid, the substrate for sialyltransferases, which disenables the cells from sialylation. This defect of an epimerase in the precursor synthesis pathway can be by-passed by addition of ManNAc to the cell culture medium in a concentration dependent manner which allows the adjustment of the sialylation degree by in process controls resulting in different sialylation glycoforms of the product. Five variants with increasing sialylation degree were analysed in a mouse model for determining the bioavailability and in *in vitro* assays for the bioactivity. As shown previously [9], the degree of sialylation clearly had a strong impact on the activity in the chosen experimental *in vitro* setting, with the highest activity at a high, but interestingly not the highest sialylation degree. The comparison of these activities with that of the commercial products revealed a manifold higher activity especially for the high but not highest sialylated isoform. When comparing the *in vivo* half-life, the optimized form with the high degree of sialylation was detectable by far for the longest time after injection in mice (Figure 4), which is in accordance to the theory that large amounts of sialic acids result in an elongated serum half life. Interestingly, the unsialylated (but nevertheless galactosylated) glycoform exhibited a half-life even shorter than that of the not-glycosylated *E. coli* product, which is assumed to be due to the large amount of free terminal galactoses.

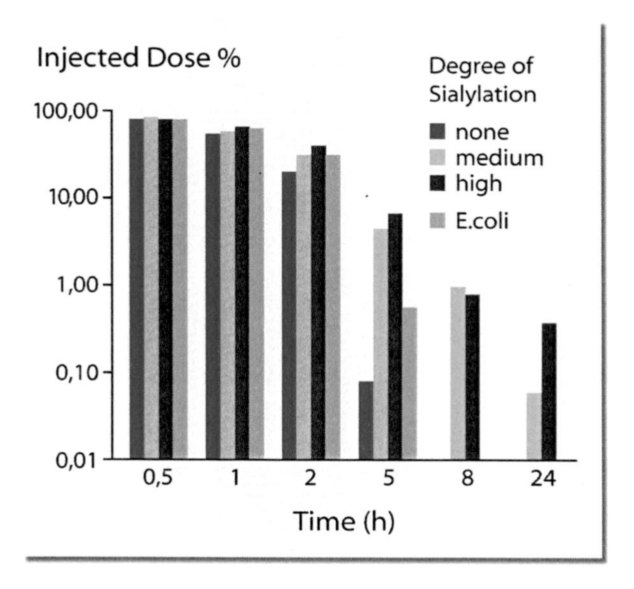

Figure 4. Improvement of the bioavailability of a glycooptimized growth factor in mice. Three glycosylation forms of one growth factor expressed in GlycoExpress (not-sialylated in blue, medium-sialylated in yellow and highly-sialylated in red) and the commercially available protein produced in *E. coli* (in grey) were injected at comparable concentrations into mice and the amount of circulating growth factor (shown in percentage of injected dose) was analysed at the indicated time after injection.

Glyco-engineered Cell Lines Chosen for Production of Therapeutic Proteins Need to Have Excellent Biotechnical Characteristics

Besides a fully human and glycooptimized glycosylation a high and competitive productivity in large scale fermentation processes are needed for pharmaceutical GMP production.

The glyco-engineered human cell lines are of leukemic origin and, by nature, suspension cells and resistant to high sheer forces. Prerequisite for biopharmaceutical production is a fully serum free system lacking any viral particles, which was proven for the various GlycoExpress cell lines by extensive RT-PCR tests, cellular reporter assays and by electron microscopy. Further advantages are the easy and rapid single cell cloning and the high transfection efficiencies as well as a very stable expression for more than 55 – 80 cell generations without selection pressure at a very rapid cell division (14 – 24 h) in high density cell cultures at outstanding cell viability rates. A proprietary gene amplification vector system and secretion signal peptides were developed for high-yield expression and an automated cell screening and cloning system was established. In consequence, GlycoExpress based production cell lines are very fast developed in 4 – 8 month with productivities of up to 45 pg of antibody/cell/day under serum-free conditions. To meet the regulatory requirements as a production cell line for biopharmaceuticals it was of advantage that the cells are of non-fetal origin, that the transformation/immortalization was non-virally (in contrast to *e.g.* Per.C 6 cells) and that the transformation events are known. It was essential that no virus or virus particle, either human or bovine, could be detected (in contrast to CHO cells) and that the history of the cell lines was documented.

A high-yield, high cell density perfusion process was established allowing over 1 – 5 g/l volumetric productivity over 14 – 20 days thereby comparable to the best CHO fed batch systems, with the ability to prolong and expand the production to several weeks. The big advantage over CHO and other fed batch systems is the uniform, highly reproducible and scalable glycosylation pattern. This was shown up to now in many non-GMP and about 10 productions under GMP conditions with up to 6000 l production volumes.

In summary, the human GlycoExpress cell lines are highly competitive in speed and productivity to the other production systems such as the most broadly used and evolved mammalian production system based on CHO cells or human Per.C 6 cells and are even superior in a whole set of characteristics. Several regulatory authorities approved the use of biopharmaceutical products produced from the GlycoExpress system for use in clinical trails in humans.

Conclusion

Glycosylation has increasingly shifted into the focus of drug development due to its potency of largely improving biopharmaceuticals in respect to bioactivity, bioavailability, immunogenicity, stability, solubility and patient coverage as shown for varies molecule types by the GlycoExpress system. Molecule classes that can be largely improved are not only antibodies but also non-antibody molecules such as protein hormones, growth factors, blood factors, cytokines, interferons, co-stimulatory factors and other glycosylated molecules covering the whole spectrum of biopharmaceuticals for humans. Today, no other technology is known to be able to improve this broad spectrum of molecules and to such an extend as shown herein in a few examples. The technology matured now to a state where production of the glycooptimized products is highly competitive and the product quality is superior, with first products in the clinic. The glycooptimized biotherapeutics are and can be further protected by new IP. Therefore, it is expected that further glycooptimized products will enter the clinic and will be of benefit for the patients.

References

[1] Varki, A. (1993) Biological roles of oligosaccharides: all of the theories are correct. *Glycobiology* **3**:97 – 130.
doi: http://dx.doi.org/10.1093/glycob/3.2.97.

[2] Kawasaki, N. *et al.* (2009) The significance of glycosylation analysis in development of biopharmaceuticals. *Biol. Pharm. Bull.* **32**:796 – 800.
doi: http://dx.doi.org/10.1248/bpb.32.796.

[3] Brooks, S.A. (2004) Appropriate glycosylation of recombinant proteins for human use: implications of choice of expression system. *Mol. Biotechnol.* **28**:241 – 255.
doi: http://dx.doi.org/10.1385/MB:28:3:241.

[4] Jenkins, N. *et al.* (1996) Getting the glycosylation right: Implications for the biotechnology industry. *Nat. Biotech.* **73**:975 – 981.
doi: http://dx.doi.org/10.1038/nbt0896-975.

[5] Noguchi, A. *et al.* (1995) Immunogenicity of *N*-glycolylneuraminic acid-containing carbohydrate chains of recombinant human erythropoietin expressed in Chinese hamster ovary cells. *J. Biochem.* **117**:59 – 62.

[6] Chung, C.H. *et al.* (2008) Cetuximab-induced anaphylaxis and IgE specific for galactose-alpha-1,3-galactose. *N. Engl. J. Med.* **358**:1109 – 1117.
doi: http://dx.doi.org/10.1056/NEJMoa074943.

[7] Shields, R.L. *et al.* (2002) Lack of Fucose on Human IgG1 N-Linked Oligosaccharide Improves Binding to Human FcγRIII and Antibody dependent Cellular Toxicity. *J. Biol. Chem.* **277**:26733 – 40.
doi: http://dx.doi.org/10.1074/jbc.M202069200.

[8] Musolino, A. *et al.* (2008) Immunoglobulin G fragment C receptor polymorphisms and clinical efficacy of trastuzumab-based therapy in patients with HER-2/neu-positive metastatic breast cancer. *J. Clin. Oncol.* **26**:1789 – 96.
doi: http://dx.doi.org/10.1200/JCO.2007.14.8957.

[9] Weng, W.-K. and Levy, R. (2003) Two immunoglobulin G fragment C receptor polymorphisms independently predict response to Rituximab in patients with follicular lymphoma. *J. Clin. Oncol.* **21**:3940 – 3947.
doi: http://dx.doi.org/10.1200/JCO.2003.05.013.

[10] Bibeau, F. *et al.* (2009) Impact of FcγRIIa-FcγRIIIa polymorphisms and *KRAS* mutations on the clinical outcome of patients with metastatic colorectal cancer treated with Cetuximab plus Irinotecan. *J. Clin. Oncol.* **27**:1122 – 1129.
doi: http://dx.doi.org/10.1200/JCO.2008.18.0463.

[11] Baumeister, H. (2006) A novel human expression system for production of higher active biotherapeutics with optimised glycosylation. *Pharma. Chem. Biopharmaceuticals* **2**:21 – 24.

Glyco-Bioinformatics – *Bits 'n' Bytes of Sugars*
October 4th – 8th, 2009, Potsdam, Germany

Structural Glycomics – Molecular Details of Protein-Carbohydrate Interactions and their Prediction

Dirk Neumann[1] and Oliver Kohlbacher[1,2,*]

[1]neumann scientific computing & consulting, Berlin, Germany

[2]Center of Bioinformatics, Division for Simulation of Biological Systems, Sand 14, University of Tübingen, Tübingen, Germany

E-Mail: *oliver.kohlbacher@uni-tuebingen.de

Received: 8th April 2010 / Published: 10th December 2010

Abstract

Protein-ligand docking is an essential technique in computer-aided drug design. While generally available docking programs work well for most drug classes, carbohydrates and carbohydrate-like compounds are often problematic for docking. We discuss the peculiarities of protein-carbohydrate interactions and their impact on protein-carbohydrate docking and review the state of the art in docking of carbohydrates to proteins. Finally, we give an overview of carbohydrate docking studies and present a new docking method specifically designed to handle docking of carbohydrate-like compounds. BALLDock/SLICK combines an evolutionary docking algorithm for flexible ligands and flexible receptor side chains with carbohydrate-specific scoring and energy functions. The scoring function has been designed to identify accurate ligand poses, while the energy function yields accurate estimates of the binding free energies of these poses. On a test set of known protein-sugar complexes we demonstrate the ability of the approach to generate correct poses for almost all of the structures and achieve very low mean errors for the predicted binding free energies.

Introduction

The interactions between carbohydrates and proteins are of special interest, because glycoproteins and glycolipids play numerous fundamental roles in biological processes such as cell fertilization, differentiation, and development. Realizing the importance of carbohydrates has even led to the term "glycomics" to indicate a new era of carbohydrate research [1].

One of the main reasons for the observed versatility of carbohydrates in transmitting information is the complex structure of carbohydrates. The huge number of possible carbohydrate structures, – glycospace – can encode huge amounts of information [2]. In mammals, however, only a small part of glycospace seems to be used [3]. This potential for transmitting information has been exploited by nature in various forms, where protein-carbohydrate interactions are crucial for many biological processes including cell-cell recognition, proliferation, and differentiation. A prominent example is the interaction of leukocyte L-selectin with its endothelial ligand, which is crucial for initiating leukocyte recruitment to sites of inflammation [4]. Glycosylation determines antibody activity, too, and is responsible for mediating triggered inflammatory responses [5 – 7]. It has also been shown that the glycosylation patterns differ between normal cells and cancer cells [8, 9].

While the protein-carbohydrate interaction may be used for defensive purposes, it has also been hijacked by other organisms, bacteria and viruses [10 – 12]. For example, *Helicobacter pylorus* binds to carbohydrates expressed on gastric epithelial cells [13]. Bacterial toxins like cholera toxin, *Escherichia coli* enterotoxin pertussis toxin, and tetanus toxin attach themselves to carbohydrates on cell surfaces and subsequently enter the cell.

Of course, the protein-carbohydrate interaction may also be exploited in novel therapeutic approaches [14, 15]. In tumour therapy for example, galectins are promising targets [16]. In the case of HIV, several proteins were reported to bind to the highly glycosylated envelope protein gp120 – an interaction which may be the basis for a vaccine [17 – 19]. More generally, protein-carbohydrate interactions may be used to enhance the efficacy of drug delivery systems [20, 21] or to define the specificity of drug targeting systems for certain, possibly malignant, cells or tissues [22, 23]. The large potential of such approaches was shown in several studies [24, 25].

A thorough understanding of the factors governing the affinity between proteins and carbohydrates might help to design tailor-made carbohydrates or proteins to better exploit this special type of interaction. However, experimental assessment of the carbohydrate-recognition by NMR spectroscopy or X-ray crystallography is impeded by difficulties of co-crystallizing proteins and carbohydrates. Here, computational modelling might help to increase our understanding of the different contributions to the binding energy. Recent developments allow searching the conformational space efficiently and yield reliable estimates of the binding free energy.

PROTEIN-CARBOHYDRATE INTERACTION

Properties of Carbohydrates

While current docking methods are quite successful when dealing with drug-like structures, they are having difficulties with carbohydrate-like structures due to their different physico-chemical properties. The average log P(octanol/water) of drugs is around 2.5. They typically have only few hydrogen bond acceptors or donors, and a small number of rotatable bonds [26, 27]. In contrast, carbohydrates and their derivatives possess many hydroxyl groups and thus a large number of rotatable bonds. Due to the many hydroxyl groups, these compounds are usually highly water soluble and their log P is often negative (Tab. 1).

Table 1. Selection of carbohydrates, carbohydrate derivatives, and drug molecules with high similarity to sugar molecules.

Compound/ Generic Name	XlogP3	#H-bond donors	#H-bond acceptors
Glucose	-2.6	5	6
Cellobiose	-4.7	8	11
4-nitrophenyl-β-D-glucoside	-0.4	4	8
Phenylglucoside	-0.9	4	6
Acarbose	-8.5	14	19
Ademoetionine	-12.2	16	49
Anthelmycin	-8.5	13	17
Digoxin	1.3	6	14
Paromomycin	-8.7	13	19
Propikacin	-8.0	13	17
Streptomycin	-8.0	12	19
Tobramycin	-6.2	10	14

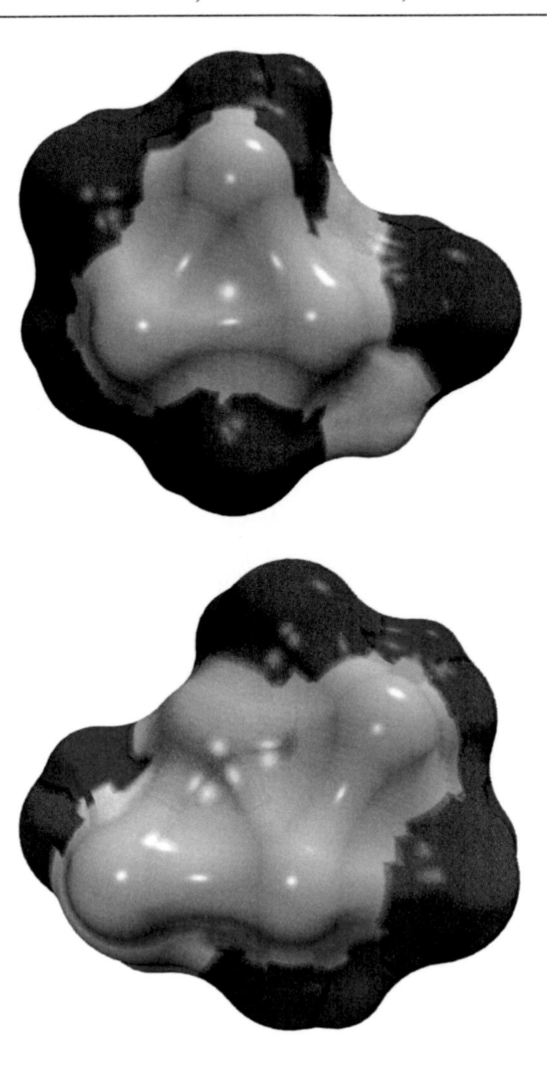

Figure 1. Electrostatic potential of β-D-glucopyranose. Color code: blue – positive charge, red – negative charge.

The surface of carbohydrates and their derivatives is composed of hydrophobic and hydrophilic patches formed by nonpolar aliphatic protons and polar hydroxyl groups (Fig. 1), which leads to anisotropic solvent densities around carbohydrate molecules [28]. In aqueous solution, favourable interaction of water molecules with the hydrophilic patches results from electrostatic interactions and hydrogen bonding. At the same time, the interaction of water with hydrophobic surface patches is unfavourable. The balance between hydrophobic and hydrophilic patches is essential for carbohydrate solubility, but also for molecular recognition.

Another essential property of carbohydrates is their conformational flexibility. Compared to drug-like molecules, carbohydrates are typically much more flexible. Similar to amino acid dimers, the energetically favourable conformations of carbohydrate dimers may be easily shown by Ramachandran plots. Figure 2 shows the Ramachandran plots for a dipeptide (with end caps) and cellobiose, a glucose dimer. At a cursory glance, both plots look very similar featuring multiple minima, with the separating energy barriers being higher in the case of cellobiose. However, carbohydrates in complex were found to adopt conformations belonging to different minima. These observations underline the necessity for thoroughly sampling the conformational space of carbohydrate oligomers during docking. While this may be feasible for glycosidic bonds, the number of degrees of freedom increases rapidly when additionally taking into account the orientation of the hydroxyl groups.

Carbohydrate Binding Proteins

The binding partners of carbohydrates encompass a wide variety of proteins including periplasmic receptor proteins, antibodies, lectins and enzymes. Lectins are proteins which bind carbohydrates with high specificity and which have been found both in plants and living organisms. While the biological roles of plant lectins are not fully understood by now, the functions of many lectins in higher organisms and especially in mammals have been elucidated in detail giving rise to the following short definition: lectins are molecular tools to decipher sugar-encoded messages [29].

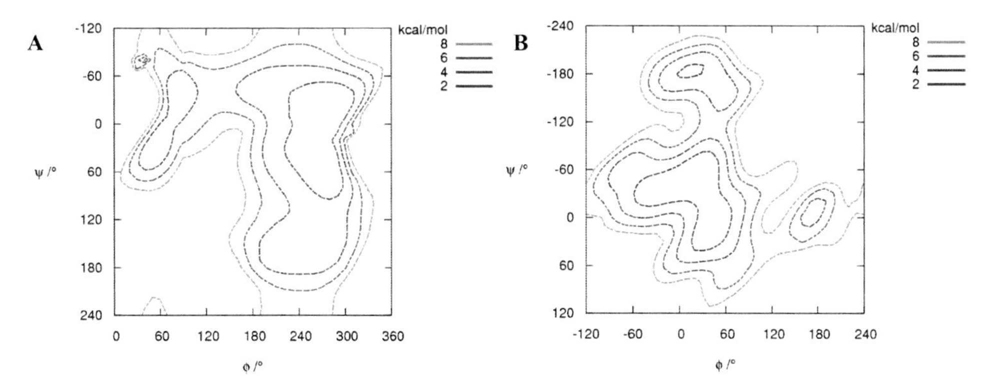

Figure 2. Comparison of peptide and carbohydrate flexibility. (A) Ramachandran plot for ACE-VAL-VALNME. (B) Energy map for cellobiose. Energies were calculated using MM3 and are relative to minimum energy found.

A variety of lectins is involved in the innate immune system: For example, the binding of the mannose-binding lectin is one of the first steps in opsonization of pathogens. Galectins, β-binding animal lectins which are involved in intracellular signalling pathways, also play important roles in immune and inflammatory responses [30]. Interestingly, galectins are not only localized in the cell membrane or intracellularly, but have been detected in extracellular space, too – very probably exerting different biological functions. The binding sites of lectins are usually rather shallow, located near the protein surface and thus accessible to solvent, whereas the binding sites of periplasmic receptors are buried in these proteins [31].

In addition to the lectins, enzymes like isomerases, glycoside hydrolases and glycosyltransferases belong to the class of carbohydrate binding proteins, too. While glycoside hydrolases (*e.g.*, lysozymes and chitinases) catalyze the hydrolysis of the glycosidic linkage, glycosyltransferases catalyze the transfer of a monosaccharide unit to, *e.g.*, a lipid or protein.

However, irrespective of its biological effect and in analogy to the surface of carbohydrates the protein surface may be divided in patches which interact favourably with water molecules and patches where rearrangement of the water molecules leads to an unfavourable interaction.

Contributions to the Binding Free Energy

Macroscopically, the formation of complexes between carbohydrates and proteins may be driven by favourable changes in enthalpy or entropy.

Usually, the binding free energies ΔG of carbohydrate monomers binding to proteins are quite small. For dimers or higher oligomers, ΔG increases significantly [32], but usually even these binding free energies are smaller than the binding free energies of other complexes such as, *e.g.*, MHC-peptide complexes [33]). In spite of the small binding free energies, the interaction between proteins and carbohydrates plays a crucial role in biological processes and may be used for pharmaceutical purposes, *e.g.* for modifying drug delivery systems with lectins, due to the polyvalent effect which leads to a high avidity [34] (for a single lectin molecule, there is a large number of possible interaction partners on a single cell surface). In addition to this obvious statistical effect, the simultaneous binding of, *e.g.*, two or more lectins with a glycocalyx may anchor a drug delivery system to a cell.

When performing docking studies, however, the variable of interest is the binding free energy between a single carbohydrate and a single carbohydrate binding protein. Usually, ΔG is assumed to be composed of independent contributions. Typically, they include hydrogen bonding, van-der-Waals forces, and consequences of the hydrophobic effect.

The exact terms used to approximate ΔG or ΔH may differ, because they inherently depend on the underlying model used for energy decomposition. In the following we will shortly outline several contributions to the binding free energy.

Hydrophobic Effect

Protein-ligand complexes may be stabilized by the so-called hydrophobic effect. In protein-carbohydrate interactions, this effect often leads to the stacking of aromatic residues against the faces of sugars. Two variations of the hydrophobic effect are discussed in literature: the "classical" hydrophobic effect and the "non-classical" hydrophobic effect [35]. The classical hydrophobic effect results mainly from a highly favourable entropy of formation. Here, small hydrophobic solutes induce an ordering of the water molecules at the solvent-surface interface. Decrease of the hydrophobic surface area upon complexation leads to a decrease in solvent ordering and hence to favourable changes in entropy. In the case of the "non-classical" hydrophobic effect, the complex formation is mainly enthalpy driven due to favourable interactions between the solute molecules forming the complex as well as favourable interactions between the solvent molecules.

Such non-classical hydrophobic effects have been reported for lectin-carbohydrate complexes [36], where the solvent interactions may provide 25 – 100% of the enthalpy of binding [37].

CH/π Interactions

The enthalpy for carbohydrates binding to proteins is characterized by a special type of interaction, the so called CH/π interaction. The CH/π interaction was defined as a hydrogen bond formed between a hydrogen attached to a carbon atom and the π system of arenes. Although much weaker than classical hydrogen bonds, CH/π interactions may contribute significantly to the enthalpy of binding. On the one hand, the presence of the electronegative oxygens bound to the carbon atoms probably increases the enthalpy resulting from the formation of a single CH/π interaction. On the other hand, multiple CH/π interactions may be formed already in the case of a methyl group interacting with a benzene. The number of interactions may be even higher for the stacking of glucose or galactose with aromatic sidechains.

The calculation of this effect using quantum mechanics is impeded by the high level of theory necessary [38]. Apart from enhancing the affinity between carbohydrate and protein, CH/π interactions probably determine the specificity of the carbohydrate binding proteins as well.

Hydrogen Bonds

Hydrogen bonds may be established between polar hydrogen atoms and lone pairs of hydrogen bond acceptors. Due to the large number of hydrogen bond donors and acceptors present in carbohydrates, they tend to form hydrogen bonds when in complex with a protein.

Here, both binding partners compete with water molecules for the hydrogen bonds. As a result, the overall enthalpic gain from a hydrogen bond formed between carbohydrate and protein may be small.

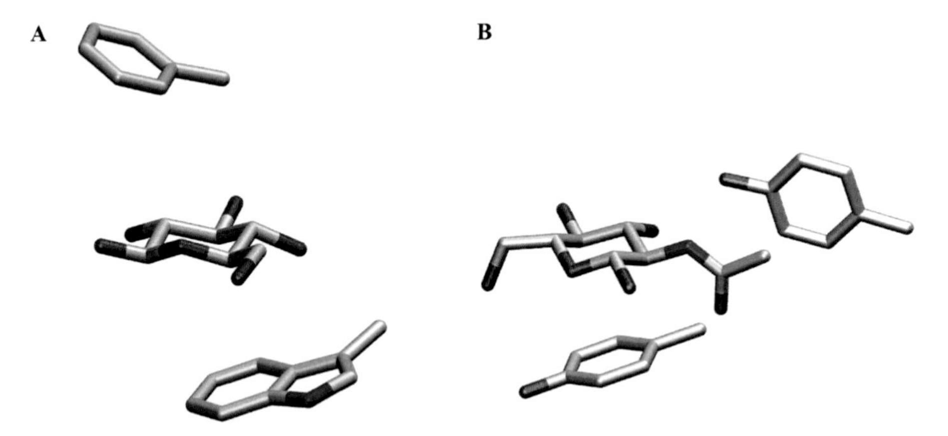

Figure 3. Stacking of (A) β-D-glucose and aromatic sidechains of the glucose/galactose binding protein (PDB accession code: 2FVY) and (B) of *N*-acetyglucosamine and tyrosine residues in wheat germ agglutinin (PDB accession code: 2UVO).

Although carbohydrates often displace all water molecules in the binding site, in a number of cases conserved water molecules are observed in the binding site. These water molecules mediate protein-carbohydrate interactions, especially if no or only few direct hydrogen bonds are established [39, 40]. However, water molecules may also help in stabilizing oligosaccharide conformations [41]. Although there exist some approaches for taking into account solvent molecules in the binding site [42], calculating water mediated hydrogen bonds of the carbohydrate with itself is still very difficult.

Electrostatic Interactions and van-der-Waals Interactions

Electrostatic interactions take place between partially charged atoms. The effects are higher for charged molecules such as sialic acids, sulfatides or protonated or deprotonated protein sidechains. Electrostatic forces are part of many force fields used in molecular modelling and thus routinely computed using Coulomb's law. The calculation of the necessary atomic partial charges of carbohydrates for use in established force fields has been repeatedly addressed in literature [43]. For docking purposes, often a reduced dielectric constant or a distance-dependent dielectric function is employed.

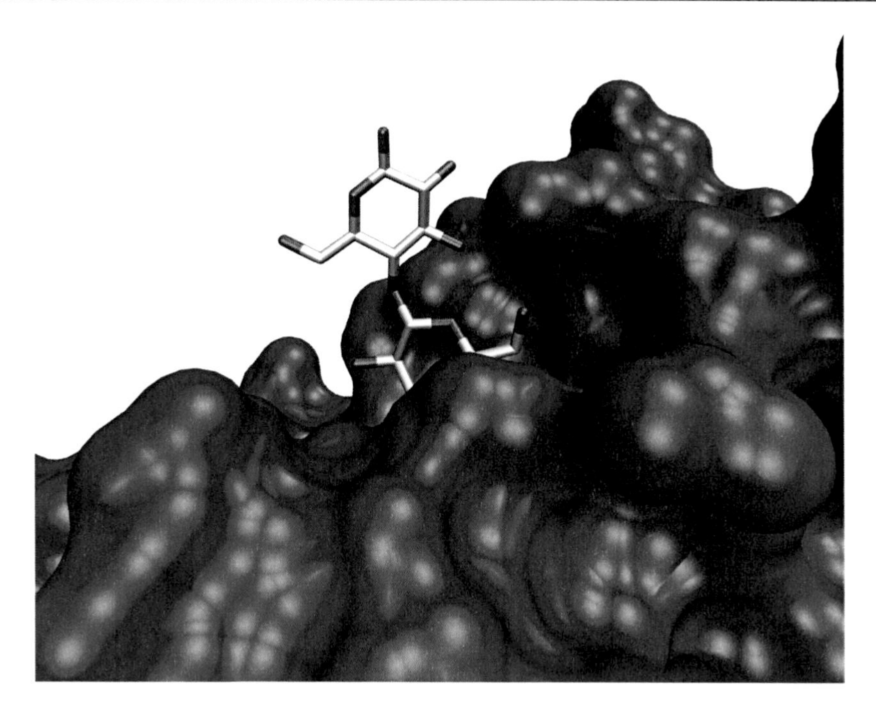

Figure 4. Example of a carbohydrate dimer extending into the solvent instead of forming a closely packed complex (PDB accession code: 1UZY).

In contrast to electrostatic interactions, van-der-Waals forces are, roughly speaking, weak attractive forces arising from the interactions between – not necessarily permanent – dipoles. At short ranges, however, overlap between electron orbitals results in a strong repulsion between atoms. The effect of both interactions on complex stability is often modelled using the Lennard-Jones potential. While the energetic contribution of a single pair of atoms to the total binding free energy may be small, summation over all pairwise interactions occurring between protein and ligand may result in a comparatively high favourable interaction.

Solvation/Desolvation

A typical problem that arises in docking carbohydrate dimers and oligomers is the tendency of docking programs to maximize atomic contacts between ligand and protein. This results in structures with small volume, where the carbohydrate lies more or less flat on the protein surface. However, several X-ray structures show contradictory structures with carbohydrate residues extending well into the surrounding solvent (Fig. 4).

Possibly, these structures might be correctly predicted, if the impact of solvation and desolvation on the binding free energy was computed with sufficient accuracy. The more accurate methods, in particular Poisson-Boltzmann methods, are unfortunately computationally rather expensive. While they can be used in re-scoring of docking conformation, their use in the inner loop of docking, the pose generation, is still prevented by large computation times.

Protein-Carbohydrate Docking

Force Fields and Scoring Functions for Carbohydrate Docking

To simulate the behaviour of carbohydrate *in vacuo* or in solution (*e.g.*, to study ring puckering [44] or rotational barriers of oligosaccharides), either established force fields or special parameterizations may be used [45 – 49]. Such force fields allow for investigating the deformation of carbohydrate rings as well as predicting. These special force fields (as well as previously established ones) have been employed repeatedly for molecular dynamics simulations of protein-carbohydrate complexes [50, 51]. In some cases, the simulations were even used for estimating successfully binding free energies [40, 52 – 54].

Despite the many possible advantages of established force fields, they were not designed to predict binding free energies or enthalpies in protein-ligand docking. Since solvent molecules are usually modelled explicitly, force fields do not need to include extra terms for hydrophobic effects. The special CH/π interactions are not taken into account. Some force fields do model for hydrogen bonds explicitly, while others regard it as part of the electrostatic interaction. Irrespective of the approach, displacement of water molecules competing for hydrogen bonds is not accounted for.

Although some force fields correlate well with *ab initio* calculations for *ab initio* optimized geometries [55], a scoring function for docking must be able to distinguish between decoy structures and true hits. A recent comparison of the results of *ab initio* and force field calculations underlines the difficulties in predicting binding enthalpies in protein-carbohydrate complexes using existing force fields: the stabilizing interaction energy for the interaction between fucose and tryptophane is heavily overestimated by the AMBER* force field [56].

Search Algorithms and Docking Programs

A major problem in docking carbohydrates is the possibly large number of degrees of freedom to be optimized during the docking calculation. In many force fields and consequently in most docking programs, polar hydrogens are modelled explicitly, which leads to many rotatable bonds and, hence, numerous degrees of freedom when docking carbohydrates. Since well-established docking algorithms are able to handle only a limited number of degrees of freedom with high efficiency, they may fail due to limited sampling of the conformational space.

Case Studies

Structures of proteins in complex with carbohydrates or carbohydrate derivatives have been available from the Protein Data Bank for some time and thus have been part of the data sets used for developing docking algorithms and docking programs.

A protein that has been repeatedly investigated in this context is neuraminidase. Native docking experiments of neuraminidase ligands performed with PRO_LEADS showed that these ligands are more difficult to dock than others [57]. The failure was attributed mainly to a failure of the energy function used: for two neuraminidase derivatives, decoy structures were ranked better than the crystal structure. However, another interesting aspect is the small size of the two top ranking clusters obtained in the docking experiments with sialic acid (PDB code: 1NSC): only 6 of the 100 docking results populate the top ranked cluster located in the binding site. For two of the complexes, experimental binding affinities are available and can be compared with predicted energies: in one case, the binding energy is over-estimated by 5 kJ/mol, whereas in the other it is underestimated by the same amount.

A complex very similar to 1NSC (pdb code: 2SIM) was successfully reproduced by the program GOLD [58]. Like PRO_LEADS, the GOLD energy function honours the formation of hydrogen bonds. Successful docking to the L-arabinose-binding protein (PDB codes: 1ABE, 1ABF) was reported repeatedly for several docking algorithms and energy/scoring functions including AutoDock 3.0.5 and MOE 2004.03 [59 – 61].

For the hemagglutinin-sialic acid complex 4HMG, however, docking using AutoDock and MOE 2004.03 failed [59]. Despite the success of AutoDock in identifying the binding pose, it overpredicted the binding free energy by more than 10 kJ/mol [62].

In comparison to AutoDock, GOLD and ICM, the docking programs FlexX and DOCK, which both use an incremental reconstruction algorithm, failed to identify the correct binding pose for some complexes of neuraminidase and L-arabinose binding protein [63]. This hints at a problem of these algorithms with selecting and placing the first ligand fragments. In a series of publications, the glucosyl saccharides were docked into the glucoamylase active site using AutoDock [64, 65]. For the 9 investigated disaccharides, the solutions in highest ranked cluster corresponded to the crystal structures. These results underlined the potential of the AutoDock program and its scoring function in docking carbohydrates.

In 1999, Minke *et al.* published a method to predict water interactions for use in carbohydrate docking studies [66]. The authors employed AutoDock 2.2 to develop their methodology using known X-ray structures of heat-labile enterotoxin. Subsequently, they performed docking calculations with a number of carbohydrate derivatives without water molecules and with predicted water molecules. In some cases, inclusion of the predicted water molecules enabled the program to better identify the binding pose in the blind docking tests. Despite the partial success, the AutoDock scoring function was not able to rank true hits better than decoy structures in all cases. In summary, the success of this scoring function seems to depend on the properties of the protein-carbohydrate complex investigated.

Since most docking programs seem to neglect the peculiar interactions in protein-carbohydrate complexes, specific scoring functions for carbohydrate docking have been developed.

Starting from the AutoDock scoring function, Reilly and co-workers identified new coefficients as well as additional terms for use in carbohydrate docking with AutoDock 3.0.6. In a first study, a training set of 30 protein-carbohydrate complexes was employed to fit the coefficients and to investigate the effect of two different hydrogen bonding terms on the calculated binding free energy. The test set consisted of 17 complexes with experimentally determined binding free energies. The smallest error in predicted binding free energies was obtained, when docking was performed using a molecular mechanics potential energy function and the results were re-evaluated using the newly developed free energy model [67]. In a subsequent study, a much larger training set was used to re-fit the parameters and to further extend the previous free energy model [68]. The best model featured a root mean squared error less than 8.5 kJ/mol. However, the authors conclude that the solvation and entropic terms are still partially poorly modelled.

Kerzmann *et al.* chose a similar approach [69, 70]. However, the authors used two different functions throughout their studies: one function for (re-)scoring putative solutions (SLICK/score) and another, much more elaborate function to calculate the binding free energies of promising docking results (SLICK/energy). Both scoring and energy function contained special terms for hydrogen bonding and CH/π interactions accounting not only for the distance-dependence but also for the directionality of these interactions (Fig. 5).

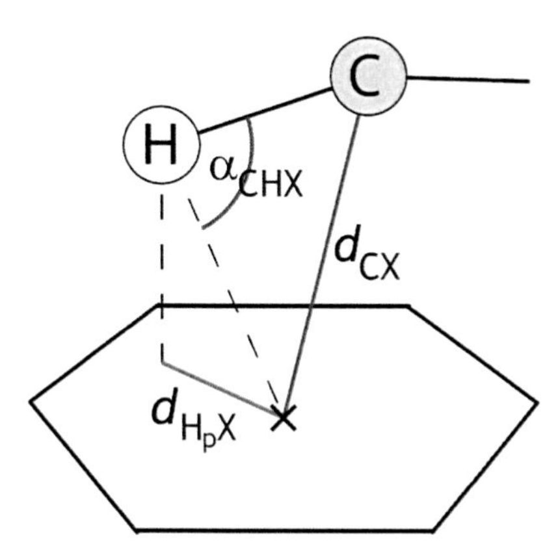

Figure 5. The geometry of the CH/π interaction employed in SLICK. d_{CX} is the distance between aliphatic carbon and ring centre, α_{CHX} denotes the angle between CH bond and the connection between the hydrogen atom and the ring centre. d_{H_pX} is the distance of H to the ring centre projected into the ring plane.

Van-der-Waals and electrostatic interactions were calculated essentially using the AMBER/ GLYCAM200a force field [71]. In addition, the energy function contained special terms to take into account changes in ΔG due to solvation/desolvation of the ligand and protein. Originally, scoring and energy function were trained on a set of 18 protein-carbohydrate complexes. Re-scoring docking results for this training set showed the potential of the new scoring function. For this purpose, the protein-carbohydrate complexes were first re-docked using AutoDock 3. Then the average rank of the first true positive, that is the first hit obtained during docking with a root-mean square deviation (RMSD) less than 1.5 Å, was calculated. The average rank for the AutoDock energy function and SLICK/score was 33.1 and 8.3, respectively. These results show that although AutoDock was able to generate solutions close to the crystal structures, its energy function ranked decoy structures higher. Using SLICK/score during docking with a genetic algorithm resulted in an average rank of the first true positive of 1.2 for the training set and 3.6 for a test set consisting of 20 lectin-carbohydrate complexes. Binding free energies were calculated with SLICK/energy for all docking results of the training set. The Spearman rank correlation coefficients were $\rho_{ftp} = 0.71$ and $\rho_{min} = 0.85$ for the energies of the first true positives and of the hits with minimal RMSD, respectively. This result shows that even small deviations from the crystal structure may lead to deterioration of the calculated binding free energy and underlines the necessity to thoroughly sample energy minima during docking. The quality of SLICK/ energy was further assessed by re-docking an external energy validation set. For these structures, the free energy model by Kerzmann *et al.* featured a root mean squared error for the binding free energy less than 4 kJ/mol. Although both training and test set used in the studies by Kerzmann *et al.* were quite small, the results are very encouraging.

In summary, in view of the good performance of the energy models presented here, the modification of existing functions for application in carbohydrate docking is certainly worthwhile. Although the number of groups active in this field is limited, the importance of protein-carbohydrate interactions in biological processes will definitely stimulate further research.

Conclusion

There exist already some promising approaches to protein-carbohydrate docking. However, to calculate binding free energies or enthalpies with the necessary accuracy, increasing the understanding of the interactions on an atomic level is crucial. Simple experimental models in conjunction with high-level *ab initio* calculations will certainly help in gaining additional insight and thus form the basis for future scoring function to be used in protein-carbohydrate docking [72]. Finally, recently established methods for handling many rotatable bonds in flexible docking will allow for docking even oligosaccharides with high efficiency [73, 74]. Ultimately, applications for carbohydrate-docking will allow for exploiting protein-carbohydrate interactions – especially for therapeutic purposes. Well-designed carbohydrate mimetics or lectinomimetics might be employed as new therapeutic agents. Similarly, the effectiveness of drug-delivery systems and the specificity of drug-targeting systems may be increased considerably.

REFERENCES

[1] Feizi, T. and Mulloy, B. (2003) Carbohydrates and glycoconjugates. Glycomics: the new era of carbohydrate biology. *Curr. Opin. Struct. Biol.* **12**:602 – 604. doi: http://dx.doi.org/10.1016/j.sbi.2003.09.001.

[2] Solís, D., Jiménez-Barbero, J., Kaltner, H., Romero, A., Siebert, H.-C., von der Lieth, C.-W. and Gabius, H.-J. (2001) Towards defining the role of glycans as hardware in information storage and transfer: Basic principles, experimental approaches and recent progress. *Cells Tissues Organs* **168**:5 – 23. doi: http://dx.doi.org/10.1159/000016802.

[3] Werz, D.B., Ranzinger, R., Herget, S., Adibekian, A., von der Lieth, C.-W., and Seeberger, P.H. (2007) Exploring the structural diversity of mammalian carbohydrates ("glycospace") by statistical databank analysis. *ACS Chem. Biol.* **2**:685 – 691. doi: http://dx.doi.org/10.1021/cb700178s.

[4] Khan, A.I., Landis, R.C. and Malhotra, R. (2003) L-selectin ligands in lymphoid tissues and models of inflammation. *Inflammation* **27**:265 – 280. doi: http://dx.doi.org/10.1023/A:1026056525755.

[5] Burton, D.R. and Dwek, R.A. (2006) Sugar determines antibody activity. *Science* **313**:627 – 628. doi: http://dx.doi.org/10.1126/science.1131712.

[6] Kaneko, Y., Nimmerjahn, F. and Ravetch, J.V. (2006) Anti-inflammatory activity of immunoglobulin G resulting from Fc sialylation. *Science* **313**:670 – 673. doi: http://dx.doi.org/10.1126/science.1129594.

[7] Altmann, F. (2007). The role of glycosylation in allergy. *Int. Arch. Allergy Immunol.* **142**:99 – 115. doi: http://dx.doi.org/10.1159/000096114.

[8] Moore, R., King, N. and Alroy, J. (1988) Differences in cellular glycoconjugates of quiescent, inflamed, and neoplastic colonic epithelium in colitis and cancer-prone tamarins. *Am. J. Pathol.* **131**:484 – 489.

[9] Gorelik, E., Galili, U. and Raz, A. (2001). On the role of cell surface carbohydrates and their binding proteins (lectins) in tumor metastasis. *Cancer Metastasis Rev.* **20**:245 – 277. doi: http://dx.doi.org/10.1023/A:1015535427597.

[10] Ofek, I. and Sharon, N. (1988) Lectinophagocytosis: a molecular mechanism of recognition between cell surface sugars and lectins in the phagocytosis of bacteria. *Infect. Immun.* **56**:539 – 547.

[11] Bomsel, M. and Alfsen, A. (2003) Entry of viruses through the epithelial barrier: pathogenic trickery. *Nat. Rev. Mol. Cell. Biol.* **4**:57 – 68.
doi: http://dx.doi.org/10.1038/nrm1005.

[12] Vigerust, D.J. and Shepherd, V.L. (2007) Virus glycosylation: role in virulence and immune interactions. *Trends Microbiol.* **15**:211 – 218.
doi: http://dx.doi.org/10.1016/j.tim.2007.03.003.

[13] Kobayashi, M., Lee, H., Nakayama, J. and Fukuda, M. (2009). Roles of gastric mucin-type O-glycans in the pathogenesis of *Helicobacter pylori* infection. *Glycobiology* **19**:453 – 461.
doi: http://dx.doi.org/10.1093/glycob/cwp004.

[14] Hong, F., Yan, J., Baran, J.T., Allendorf, D.J., Hansen, R.D., Ostroff, G.R., Xing, P.X., Cheung, N.-K.V. and Ross, G.D. (2004) Mechanism by which orally administered β-1,3-glucans enhance the tumoricidal activity of antitumor monoclonal antobodies in murine tumor models. *J. Immunol.* **173**:797 – 806.

[15] Oppenheimer, S.B., Alvarez, M., Nnoli, J. (2008) Carbohydrate-based experimental therapeutics for cancer, HIV/aids and other diseases. *Acta Histochem.* **110**:6 – 13, 2008.
doi: http://dx.doi.org/10.1016/j.acthis.2007.08.003.

[16] Ingrassia, L., Camby, I., Lefranc, F., Mathieu, V., Nshimyumukiza, P., Darro, D. and Kiss, R. (2006) Anti-galectin compounds as potential anti-cancer drugs. *Curr. Med. Chem.* **13**:3513 – 3527.
doi: http://dx.doi.org/10.2174/092986706779026219.

[17] Ji, X., Gewurz, H. and Spear, G.T. (2005) Mannose binding lectin (MBL) and HIV. Mol. *Immunol.* **42**:145 – 152.
doi: http://dx.doi.org/10.1016/j.molimm.2004.06.015.

[18] Balzarini, J. (2006) Inhibition of HIV entry by carbohydrate-binding proteins. *Antivir. Res.* **71**:237 – 247, 2006.
doi: http://dx.doi.org/10.1016/j.antiviral.2006.02.004.

[19] Sirois, S., Touaibia, M., Chou, K.C. and Roy, R. (2007) Glycosylation of HIV-1 gp120 V3 loop: towards the rational design of a synthetic carbohydrate vaccine. *Curr. Med. Chem.* **14**:3232 – 3242.
doi: http://dx.doi.org/10.2174/092986707782793826.

[20] Clark, M.A., Hirst, B.H. and Jepson, M.A. (2000) Lectin-mediated mucosal delivery of drugs and microparticles. *Adv. Drug Deliv. Rev.* **43**:207 – 223.
doi: http://dx.doi.org/10.2174/092986707782793826.

[21] Russell-Jones, G.J. (2001) The potential use of receptor-mediated endocytosis for oral drug delivery. *Adv. Drug Deliv. Rev.* **46**:59 – 73.
doi: http://dx.doi.org/10.1016/S0169-409X(00)00127-7.

[22] Minko, T. (2004) Drug targeting to the colon woth lectins and neoglycoconjugates. *Adv. Drug. Deliv. Rev.* **56**:491 – 509.
doi: http://dx.doi.org/10.1016/j.addr.2003.10.017.

[23] Nishikawa, M. (2005) Development of cell-specific targeting systems for drugs and genes. *Biol. Pharm. Bull.* **28**:195 – 200.
doi: http://dx.doi.org/10.1248/bpb.28.195.

[24] Osborn, H.M.I, Evans, P.G., Gemmell, N. and Osborne, S.D. (2004) Carbohydrate-based therapeutics. *J. Pharm. Pharmacol.* **56**:691 – 702.
doi: http://dx.doi.org/10.1211/0022357023619.

[25] McReynolds, K.D. and Gervay-Hagues, J. (2007) Chemotherapeutic interventions targeting HIV interactions with host-associated carbohydrates. *Chem. Rev.* **107**:1533 – 1552. 107:1533 – 1552.
doi: http://dx.doi.org/10.1021/cr0502652.

[26] Ghose, A.K., Viswanadhan, V.N. and Wendoloski, J.J. (1999) A knowledge-based approach in designing combinatorial or medicinal chemistry libraries for drug discovery. 1. A qualitative and quantitative characterization of known drug databases. *J. Comb. Chem.* **1**:55 – 68.
doi: http://dx.doi.org/10.1021/cc9800071.

[27] Lipinski, C.A., Lombardo, F., Dominy, B.W. and Feeney, P.J. (2001) Experimental and computational approaches to estimate solubility and permeability in drug discovery and development settings. *Adv. Drug Deliv. Rev.* **46**:3 – 26.
doi: http://dx.doi.org/10.1016/S0169-409X(00)00129-0.

[28] Liu, Q. and Brady, J.W. (1996) Anisotropic solvent structuring in aqueous solutions. *J. Am. Chem. Soc.* **118**:12276 – 12286.
doi: http://dx.doi.org/10.1021/ja962108d.

[29] Gabius, H.J., Siebert, H.C., André, S., Jiménez-Barbero, J. and Rüdiger, H. (2004) Chemical biology of the sugar code. *ChemBioChem* **5**:740 – 764.
doi: http://dx.doi.org/10.1002/cbic.200300753.

[30] Liu, F.T. and Rabinovich, G.A. (2010) Galectins: regulators of acute and chronic inflammation. *Ann. N.Y. Acad. Sci.* **1183**:158 – 82.
doi: http://dx.doi.org/10.1111/j.1749-6632.2009.05131.x.

[31] Vyas, N.K. (1991) Atomic features of protein-carbohydrate interactions. *Curr. Opin. Struct. Biol.* **1**:732 – 740.
doi: http://dx.doi.org/10.1016/0959-440X(91)90172-P.

[32] Bains, G., Lee, R.T., Lee, Y.C. and Freire, E. (1992) Microcalorimetric study of wheat germ agglutinin binding to *N*-acetylglucosamine and its oligomers. *Biochemistry* **31**:12624 – 12628.
doi: http://dx.doi.org/10.1021/bi00165a012.

[33] Logean, A., Sette, A. and Rognan, D. (2001) Customized versus universal scoring functions: application to class I MHC-peptide binding free energy predictions. *Biorg. Med. Chem. Lett.* **11**:675 – 679.
doi: http://dx.doi.org/10.1016/S0960-894X(01)00021-X.

[34] Mammen, M., Choi, S.-K. and Whitesides, G.M. (1998) Polyvalent interactions in biological systems: Implications for design and use of multivalent ligands and inhibitors. *Angew. Chemie Int. Ed.* **37**(20):2754 – 2794.
doi: http://dx.doi.org/10.1002/(SICI)1521-3773(19981102)37:20<2754::AID-ANIE2754>3.3.CO;2-V.

[35] Meyer, E.A., Castellano, R.K. and Diederich, F. (2003) Interactions with aromatic rings in chemical and biological recognition. *Angew. Chem. Int. Ed.* **42**:1210 – 1250.
doi: http://dx.doi.org/10.1002/anie.200390319.

[36] Poveda, A., Asensio, J.L., Espinosa, J.F., Martin-Pastor, M., Cañada, J. and Jiménez-Barbero, J. (1997) Applications of nuclear magnetic resonance spectroscopy and molecular modeling to the study of proteincarbohydrate interactions. *J. Mol. Graph. Model.* **15**:9 – 17.
doi: http://dx.doi.org/10.1016/S1093-3263(97)00012-0.

[37] Chervenak, M. and Toone, E. (1994) A direct measure of the contribution of solvent reorganization to the enthalpy of ligand binding. *J. Am. Chem. Soc.* **116**:10533 – 10539.
doi: http://dx.doi.org/10.1021/ja00102a021.

[38] Spiwok, V., Lipovová, P., Skálová, T., Buchtelová, E., Hašek; J. and Králová, B. (2004) Role of CH/pi interactions in substrate binding by *Escherichia coli* β-galactosidase. *Carbohydr. Res.* **339**:2275 – 2280.
doi: http://dx.doi.org/10.1016/j.carres.2004.06.016.

[39] Bourne, Y., Rougé, P. and Cambillau, C. (1990) X-ray structure of a (α-Man(1 – 3)β-Man(1 – 4)GlcNAc)-lectin complex at 2.1 -Å resolution. *J. Biol. Chem.* **265**:18161 – 18165.

[40] Clarke, C., Woods, R.J., Gluska, J., Cooper, A., Nutley, M.A. and Boons, G.-J. (2001) Involvement of water in carbohydrate-protein binding. *J. Am. Chem. Soc.* **123**:12238 – 12247.
doi: http://dx.doi.org/10.1021/ja004315q.

[41] Çarçabal, P., Jockusch, R.A., Hünig, I., Snoek, L.C., Kroemer, R.T., Davis, B.G., Gamblin, D.P., Compagnon, I., Oomens, J. and Simons, J.P. (2005) Hydrogen bonding and cooperativity in isolated and hydrated sugars: Mannose, galactose, glucose, and lactose. *J. Am. Chem. Soc.* **127**:11414 – 11425.
doi: http://dx.doi.org/10.1021/ja0518575.

[42] Rarey, M., Kramer, B. and Lengauer, T (1999) The particle concept: placing discrete water molecules during protein-ligand docking predictions. *Proteins* **34**:17 – 28.
doi: http://dx.doi.org/10.1002/(SICI)1097-0134(19990101)34:1<17::AID-PROT3>3.0.CO;2-1.

[43] Woods, R.J. and Chappelle, R. (2000) Restrained electrostatic potential atomic partial charges for condensedphase simulations of carbohydrates. *J. Mol. Struct. (Theochem)* **527**:149 – 156.
doi: http://dx.doi.org/10.1016/S0166-1280(00)00487-5.

[44] Dowd, M.K., French, A.D. and Reilly, P.J. (1994) Modeling of aldopyranosyl ring puckering with mm3(92). *Carbohydr. Res.* **264**:1 – 19.
doi: http://dx.doi.org/10.1016/0008-6215(94)00185-5.

[45] Homans, S:W. (1990) A molecular mechanical force field for the conformational analysis of oligosaccharides: comparison of theoretical and crystal structures of Manα1 – 3Manβ1 – 4GlcNAc. *Biochemistry* **29**:9110 – 9118.
doi: http://dx.doi.org/10.1021/bi00491a003.

[46] Woods, R.J., Dwek, R.A. and Edge, C.J. (1995) Molecular mechanical and molecular dynamical simulations of glycoproteins and oligosaccharides. 1. GLYCAM_93 parameter development. *J. Phys. Chem.* **99**:3832 – 3846.
doi: http://dx.doi.org/10.1021/j100011a061.

[47] Glennon, T.M. and Merz, K.M., Jr. (1997) A carbohydrate force field for AMBER and its application to the study of saccharide to surface adsorption. *J. Mol. Struct. (Theochem)* **395**:157 – 171.
doi: http://dx.doi.org/10.1016/S0166-1280(96)04949-4.

[48] Momany, F.A. and Willett, J.L. (2000) Computational studies on carbohydrates: in vacuo studies using a revised AMBER force field, AMB99C, designed for α-(1 $\rightarrow$ 4) linkages. *Carbohydr. Res.* **326**:194 – 209.
doi: http://dx.doi.org/10.1016/S0008-6215(00)00042-2.

[49] Kirschner, K.N. Youngye, A.B., Tschampel, S.M. González-Outeiriño, J., Daniels, C.R. Foley, B.L. and Woods, R.J. (2008) GLYCAM06: a generalizable biomolecular force field. carboydrates. *J. Comput. Chem.* **29**:622 – 655, 2008. doi: http://dx.doi.org/10.1002/jcc.20820.

[50] Bradbrook, G.M., Forshaw, J.R. and Pérez, S. (2000) Structure/thermodynamics relationships of lectinsaccharide complexes. the *Erythrina corallodendron* case. *Eur. J. Biochem.* **267**:4545 – 4555. doi: http://dx.doi.org/10.1046/j.1432-1327.2000.01505.x.

[51] Tempel, W., Tschampel, S. and Woods, R.J. (2002) The xenograft antigen bound to *Griffonia simplicifolia* lectin 1-B_4. *J. Biol. Chem.* **8**:6615 – 6621. doi: http://dx.doi.org/10.1074/jbc.M109919200.

[52] Liang, G., Schmidt, R.K., Yu, H.-A., Cumming, D.A. and Brady, J.W. (1996) Free energy simulation studies of the binding specificity of mannose binding protein. *J. Phys. Chem.* **100**:2528 – 2534. doi: http://dx.doi.org/10.1021/jp952911e.

[53] Pathiaseril, A. and Woods, R.J. (2000) Relative energies of binding for antibody–carbohydrate-antigen complexes computed from free-energy simulations. *J. Am. Chem. Soc.* **122**:331 – 338. doi: http://dx.doi.org/10.1021/ja9914994.

[54] Bryce, R.A., Jillier, I.H. and Naismith, J.H. (2001) Carbohydrate-protein recognition: molecular dynamics simulations and free enrgy analysis of oligosaccharide binding to concanavalin A. *Biophys. J.* **81**:1373 – 1388. doi: http://dx.doi.org/10.1016/S0006-3495(01)75793-1.

[55] Spiwok, V., Lipovová, P., Skálová, T., Vondráčková, E., Dohnálek, J., Hašek, J. and Králová, B. (2006) Modelling of carbohydrate-aromatic interactions: *ab initio* energetics and force field performance. *J. Comput.-Aided. Mol. Des.* **19**:887 – 901. doi: http://dx.doi.org/10.1007/s10822-005-9033-z.

[56] Vandenbussche, S., Díaz, D., Fernández-Alonso, M.C., Pan, W., Vincent, S.P., Cuevas, G., Cañada, F.J. Jiménez-Barbero, J. and Bartik, K. (2008) Aromatic-carbohydrate interactions: an NMR and computational study of model systems. *Chemistry* **14**:7570 – 7578. doi: http://dx.doi.org/10.1002/chem.200800247.

[57] Murray, C.W., Baxter, C.A. and Frenkel, A.D. (1999) The sensitivity of the results of molecular docking to induced fit effects: application to thrombin, thermolysin and neuraminidase. *J. Comput.-Aided Mol. Des.* **13**:547 – 562. doi: http://dx.doi.org/10.1023/A:1008015827877.

[58] Jones, G., Willett, P., Glen, R.C., Leach, A.R. and Taylor, R. (1997) Development and validation of a genetic algorithm for flexible docking. *J. Mol. Biol.* **267**:727 – 748.

[59] Wiley, E.A., MacDonald, M., Lambropulos, A., Harriman, D.J. and Deslongchamps, G. (2006) LGA/EM-Dock – exploring Lamarckian genetic algorithms and energy-based local search for ligand-receptor docking. *Can. J. Chem.* **84**:384 – 391. doi: http://dx.doi.org/10.1139/V06-012.

[60] Thomsen, R. and Christensen, M.H. (2006) MolDock: a new technique for high-accuracy molecular docking. *J. Med. Chem.* **49**:3315 – 3321. doi: http://dx.doi.org/10.1021/jm051197e.

[61] Chen, H.M., Liu, B.F., Huang, H.L., Hwang, S.F. and Ho, S.Y. (2007) SODOCK: swarm optimization for highly flexible protein-ligand docking. *J. Comput. Chem.* **28**:612 – 623. doi: http://dx.doi.org/10.1002/jcc.20542.

[62] Morris, G.M., Goodsell, D.S., Halliday, R.S., Huey, R., Hart, W.E., Belew, R.K. and Olson, A.J. (1998) Automated docking using a Lamarckian genetic algorithm and an empirical binding free energy function. *J. Comput. Chem.* **19**:1639 – 1662. doi: http://dx.doi.org/10.1002/(SICI)1096-987X(19981115)19:14<1639::AID-JCC10 >3.0.CO;2-B.

[63] Bursulaya, B.D., Totrov, M., Abagyan, R. and Brooks, C.L. III. (2003) Comparative study of several algorithms for flexible ligand docking. *J. Comput.-Aided. Mol. Des.* **17**:755 – 763. doi: http://dx.doi.org/10.1023/B:JCAM.0000017496.76572.6f.

[64] Coutinho, P.M., Dowd, M.K. and Reilly, P.J. (1997) Automazed docking of glucosyl disaccharides in the glucoamylase active site. *Proteins* **28**:162 – 173, 1997. doi: http://dx.doi.org/10.1002/(SICI)1097-0134(199706)28:2<162::AID-PROT5>3.3. CO;2-0.

[65] Coutinho, P.M., Dowd, M.K. and Reilly, P.J.(1998) Automazed docking of α-(1,4)- and α-(1,6)-linked glucosyl trisaccharides in the glucoamylase active site. *Ind. Eng. Chem. Res.* **37**:2148 – 2157. doi: http://dx.doi.org/10.1021/ie9706976.

[66] Minke, W.E., Diller, D.J., Hol, W.G. and Verlinde, C.L. (1999) The role of waters in docking strategies with incremental flexibility for carbohydrate derivatives. *J. Med. Chem.* **42**:1778 – 1788. doi: http://dx.doi.org/10.1021/jm980472c.

[67] Laederach, A. and Reilly, P.J. (2003) Specific empirical free energy function for automated docking of carbohydrates to proteins. *J. Comput. Chem.* **24**:1748 – 1757. doi: http://dx.doi.org/10.1002/jcc.10288.

[68] Hill, A.D. and Reilly, P.J. (2008) A Gibbs free energy correlation for automated docking of carbohydrates. *J. Comput. Chem.* **29**:1131 – 1141. doi: http://dx.doi.org/10.1002/jcc.20873.

[69] Kerzmann, A., Neumann, D. and Kohlbacher, O. (2006) SLICK – scoring and energy functions for proteincarbohydrate interactions. *J. Chem. Inf. Model.* **46**:1635 – 1642. doi: http://dx.doi.org/10.1021/ci050422y.

[70] Kerzmann, A., Fuhrmann, J., Kohlbacher, O. and Neumann, D (2008) BALLDock/SLICK: a new method for protein-carbohydrate docking. *J. Chem. Inf. Model.* **48**:1616 – 1625. doi: http://dx.doi.org/10.1021/ci800103u.

[71] Cornell, W.D., Cieplak, P., Bayly, C.I., Gould, I.R., Merz, K.M., Ferguson, D.M., Spellmeyer, D.C., Fox, T., Caldwell, J.W. and Kollman, P.A. (1995) A second generation force field for the simulation of proteins, nucleic acids, and organic molecules. *J. Am. Chem. Soc.* **117**:5179 – 5197. doi: http://dx.doi.org/10.1021/ja00124a002.

[72] Terraneo, G., Potenza, D., Canales, A., Jiménez-Barbero, J., Baldrige, K.K. and Bernardi, A. (2007) A simple model system for the study of carbohydrate-aromatic interactions. *J. Am. Chem. Soc.* **129**:2890 – 2900. doi: http://dx.doi.org/10.1021/ja066633g.

[73] Trott, O. and Olson, A.J. (2009) AutoDock Vina: Improving the speed and accuracy of docking with a new scoring function, efficient optimization, and multithreading. *J. Comput. Chem.* **31**:455 – 461.

[74] Fuhrmann, J., Lenhof, H.P., Rurainski, A. and Neumann, D. (2010) A new Lamarckian genetic algorithm for flexible ligand-receptor docking. *J. Comput. Chem.*, page in press.

Software Tools for Storing, Processing and Displaying Carbohydrate Microarray Data

Mark Stoll* and Ten Feizi#

The Glycosciences Laboratory, Imperial College London, Northwick Park Campus, Watford Road, Harrow, Middlesex HA1 3UJ, U.K.

E-Mail: *m.stoll@imperial.ac.uk, #t.feizi@imperial.ac.uk

Received: 1st March 2010 / Published: 10th December 2010

Abstract

We describe a suite of software modules to store, retrieve and display carbohydrate microarray data. Storage is in a relational database that holds all the microarray data and associated glycan, protein and experimental information. The retrieval and display software has a comprehensive system of sorters, filters and arrangers to allow highly customized presentation of data as charts, tables, 2D matrices and array graphics. Matrices allow arrangement of proteins in one axis and glycans in the other, so that comparisons can be made between the binding patterns of proteins. Sorting and filtering includes a large assortment of built-in parameters that range from glycan features to data grouping in slides and experiments but may also be completely customized to suit individual needs. Charts, tables and matrices are customizable to maximize presentation clarity. There are customizable automatic chart titles, chart axis annotation and scaling, table layouts, matrix arrangements and colour schemes for all graphics. All display output from the software can be saved or printed for permanent record.

Background

Over the past 25 years we have developed and have been using the neoglycolipid (NGL) technology for studying the interactions of glycan probes and carbohydrate binding proteins. The lipid linked probes generated have been invaluable for the discovery of novel glycan ligands, many derived from natural sources and available in only minute amounts. Our laboratory has developed chromatographic and mass spectrometric methods for determining the sequences of the glycan ligands in their NGL forms [1 – 4].

Emergence of carbohydrate microarrays as powerful tools in biology and biochemistry

After gene and protein microarrays, carbohydrate microarrays have emerged and they are revolutionizing studies of carbohydrate-protein interactions, which are of fundamental biological importance in endogenous recognition systems and pathogen host interactions [5 – 8]. The advantages of microarrays over the conventional approach are parallel measurements of interactions involving thousands of samples using femtomole amounts of glycan probe. The NGL-based carbohydrate microarray system established in our laboratory at Imperial College is currently one of the two most comprehensive internationally and it includes a diverse repertoire of natural and synthetic saccharide probes (Figure 1), the other system being that of The Consortium for Functional Glycomics (CFG).

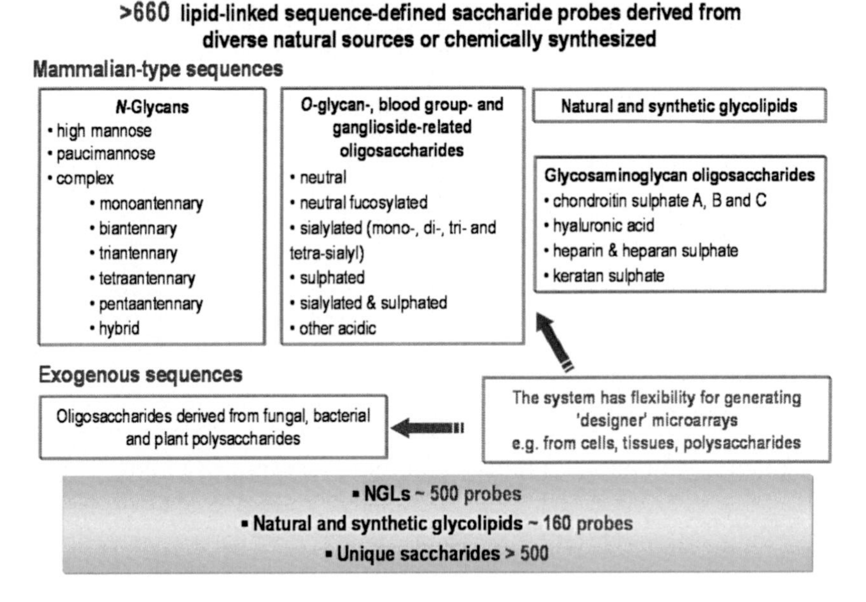

Figure 1. The Glycosciences Laboratory's repertoire of lipid-linked probes. The figure shows the range of different glycans: N-glycans, O-glycans, blood group- and ganglioside-related glycans, natural and synthetic glycolipids, glycosaminoglycans and glycans derived from fungi and bacteria.

Software for microarray data handling and storage

Software is required for carbohydrate microarray data analysis but the commercially available software packages for microarray analyses are designed for gene microarrays. The requirements for carbohydrate microarrays are very different because of the wide range of structural elements present in glycans. These include not only the different monosaccharide units and their substituents, *e.g.* sulphation, phosphorylation and acetylation, but also the linkages and anomeric configurations of the monosaccharides, branching patterns, structural motifs *e.g.* N-glycans, O-glycans, glycosaminoglycans, and presence of antigenic groupings *e.g.* blood group markers. Data formats for glycan storage are being developed by EuroCarbDB, www.eurocarbdb.org (*e.g.* GlycoCT and LINUCS), CFG, www.functionalglycomics.org (*e.g.* GLYDE-II, IUPAC format) and Kyoto Encyclopedia of Genes and Genomes (KEGG, www.genome.jp/kegg *e.g.* KCF) to address the problem of glycan structural diversity. There is software available for assignments of structure from mass spectrometric (MS) analyses and for automatic annotation of MS spectra [9, 10]. There is also software for manipulation of glycan structure and for conversion to and from cartoon representations of glycans to storage formats such as GlycoCT and LINUCS (EuroCarbDB website), Glyde and IUPAC (CFG website, Links,) and KEGG. A good summary of the present state of glyco-bioinformatics is available [11]. Currently, there is no generally available software dedicated to the analysis, storage and presentation of microarray data at laboratory level of sufficient flexibility to be of general use to any scientist wishing to work in the field.

EuroCarbDB, KEGG and CFG have large on-line database resources in which glycans, proteins and microarray screening data are linked through several databases. A customizable stand-alone software suite that can be tailored to the exact needs of individual research laboratories is desirable. Data stored in such a stand-alone system could potentially be formatted for, linked to and included in international resources.

We have been developing a suite of specialized software tools to support our carbohydrate microarray screening program. This software is in constant use in our laboratory and is being continually modified and tailored to requirements as they arise. We find it an essential adjunct to the biochemical work to help with interpretations of binding specificities.

Microarray Software

In the early stages of development, processing of our microarray data was carried out manually using the (Microsoft) Excel spreadsheet. Obtaining graphical and tabular output from a single slide of data would often take days of work. Moreover the lack of flexibility meant that changes in the way information was presented involved much further work. There was clearly a need for software that could radically improve both throughput and flexibility in data handling. Custom designed software would provide both speed and flexibility and would allow analysis and presentation of data to our exact requirements.

Furthermore, by programming in-house, we are able to adjust the software to be in line with developments in the microarray technology and meet the ever-changing requirements for processing and presentation.

Rationale for software development

For rapid and flexible development, an integrated development environment was used, namely Microsoft Office. The Office suite is very familiar to most scientists both specialists and non-specialists so that the learning curve in using the software is relatively shallow. The use of the built-in Visual Basic for Applications (VBA) using the Office Object model and the use of VBA-wrapped Windows API calls for graphics routines has proved wholly adequate in terms of flexibility and performance. All data, including user-defined preferences, are stored in an ACCESS database (DB). EXCEL-based software both transfers data to the DB and retrieves data from it for analysis and display and both EXCEL and WORD are used to store tabular and graphical output. The flow of information is indicated in Figure 2.

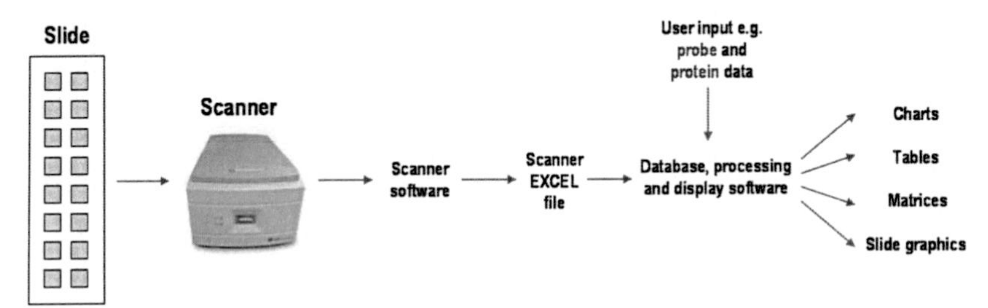

Figure 2. The flow of data from the microarray slide, scanner in EXCEL format, integration with probe and protein data via the input software, storage in an ACCESS database, retrieval and processing with the output software to give charts, tables, matrices and slide graphics.

Microarray format

The NGL microarray technology currently in use in our laboratory uses a screening format in which each glycan probe is used at two levels (2 and 7 (or 5) fmol) each in duplicate. Two fluorescent dyes are used, one (Cy3) to indicate positions of arrayed spots and the other (Alexa 647) to measure binding signals. Each microarray slide typically has the same 64 probes on the 16 nitrocellulose coated pads or two different sets of 64 probes on the two columns of pads. Thus 16 or 8 proteins, respectively, are overlaid for analysis of binding. Sixteen pads each with 64 glycan probes yields 8192 fluorescence signals (Figure 3). Over the past few years our laboratory has accumulated data from over 750 slides (>9000 binding experiments). To cross-relate and interpret this data has required the development of specialized software for storage, analysis and display.

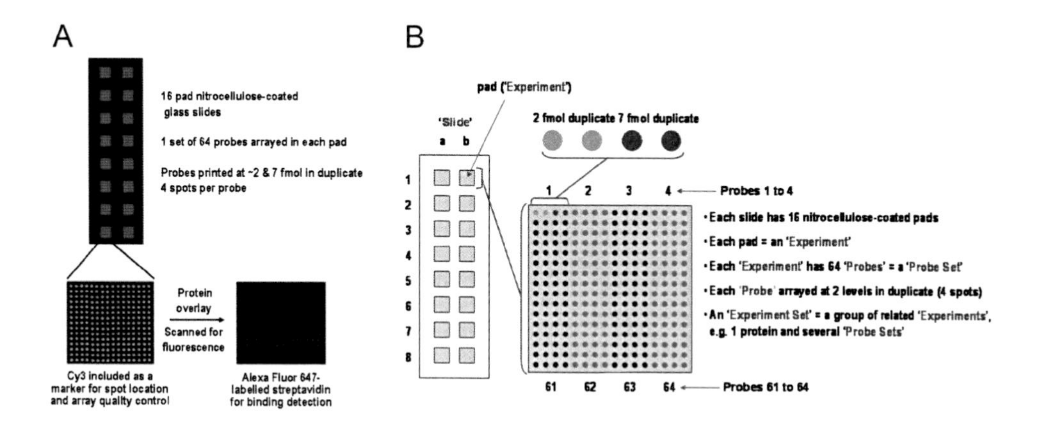

Figure 3. (A) Scanner images of a whole slide showing the Cy3 fluorescence of arrayed spots and a single experiment (nitrocellulose coated pad) showing both Cy3 fluorescence and the Alexa Fluor 647 binding signals. (B) A diagrammatic representation of a microarray slide showing the layout of probes in each experiment and the way 16 experiments are laid out in a slide. The arrangement of probes in 'Probe Sets' and 'Experiments' in 'Experiment Sets' is explained. The scanner reads the microarray slide left to right and top to bottom, one experiment at a time.

Data Input Software

Raw microarray data are returned by the scanner in EXCEL spreadsheet format. Each scanner file contains one slide of data but at between one and three laser power levels. Different laser powers are used to find optimum scan conditions to minimise noise for low level signals while avoiding saturation for high levels.

Data input

There is a graphical user interface (GUI) with an interactive slide graphic representing each experiment (16 per slide), to allow transfer of data to the DB from the laser scanner output software, together with associated data on the carbohydrate probes, carbohydrate binding proteins and experimental conditions, which may already be present in the DB associated with other microarray data or which may be added through the input interface as required (Figure 4).

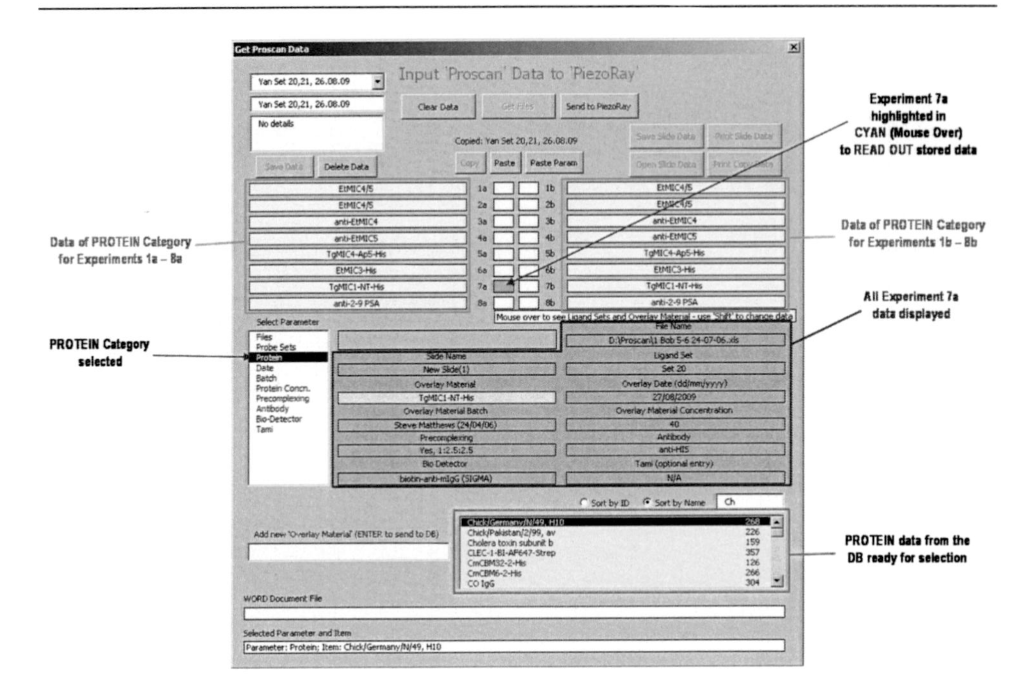

Figure 4. The user-interface of the data input software. The slide graphic is interactive and the mouse can be used to select each experiment either for data input or readout. Each parameter, when selected from the list of categories, reveals the items in the database for that parameter. Items can be selected from the list to associate with each experiment. When all the associations have been made the data are saved in the database.

Data storage

One DB holds all the data on the microarray experiments, carbohydrate probes, carbohydrate binding proteins and experimental conditions. In the DB all microarray data are stored in one table with references to all associated data stored in other tables. Each laser power level is treated as a separate slide of data for storage purposes. To maintain referential integrity the storage process is complex, involving two-way communication between the input software and the DB, which must generate unique reference identifiers for each new 'Slide' and 'Experiment'. Within the main microarray data table each record holds the four Cy3 and four Alexa Fluor 647 fluorescence values for one Glycan Probe/Protein interaction (four microarray spots) together with a slide position (1 – 1024; one for each glycan probe) and a reference each to an 'Experiment' and a Set of 64 Glycan Probes ('Probe Set'). Thus each slide is stored as 1024 records (16 experiments) with 64 records to each 'Experiment'. Glycan Probe data are stored as a text-based structure representation with both structural and immunological data stored in fields that can be used for sorting and filtering. At present we have not adopted *e.g.* the GlycoCT or GLYDE format for storing glycan information but intend to do this in future when standards are more established. In the DB, Glycan Probes are associated in sets of 64 per 'Experiment' in a separate table and 'Experiments' are grouped as 'Experiment Sets' in yet another table. These associations are made by the user. Other tables hold protein and experimental conditions data all referenced to each 'Experiment'. User-defined preferences are compressed and stored in special tables as custom character delimited strings. These handle custom sorters and filters, colour schemes, configuration settings etc. When required they are unpacked and used by the software. We use a central comprehensive DB but also smaller individual DBs for trial experiments that may or may not be transferred to the central DB.

Data retrieval

Our data analysis and presentation software (Figure 5) allows retrieval of microarray data from the DB in one of 5 formats; 'Experiment', 'Experiment Set', 'Slide', 'Glycan Probe' or 'Protein'. The full content of the DB under each category is presented to the user as a list at program start-up. Any number of items from any one category can be requested at one time to become the current data to work with. There is also an interactive 'Find' feature that allows very specific data to be retrieved in 'Experiment' format.

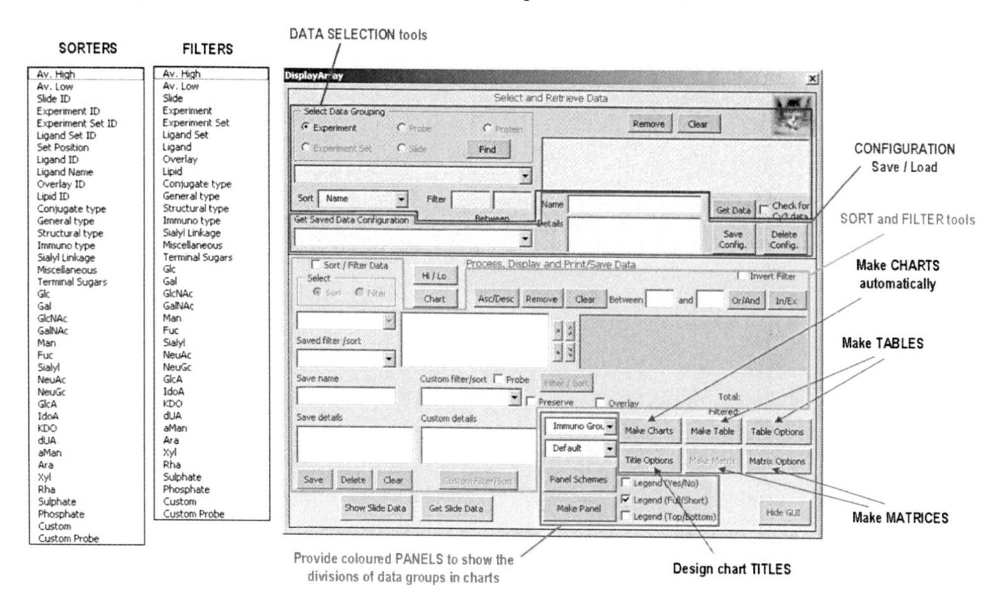

Figure 5. The user-interface of the data retrieval, processing, and presentation software. There are tools for data selection in one of 5 formats, for sorting and filtering data with a wide range of built-in as well as customizable parameters, for making tables, charts and matrices, for saving the state of the software to return to and for viewing data in slide graphic format. (The term 'ligand' refers to 'glycan probe').

The selected data are retrieved by using a complex SQL query so that all the fields (including calculated fields) that may be required by the software are available all at once. The returned Recordset Object (RO) is cloned into a Stream Object so that the data is held separately from the DB. A Collection Object also holds a number of smaller Recordsets used to populate lists used extensively by different elements of the software.

Data analysis and presentation

Charts: Charts are a major item in our data analysis and presentation and so much effort has gone into making these as versatile as possible. The fields used for building a histogram chart are copied from the RO to a hidden EXCEL worksheet linked to the software's interactive Chart Object and thus all the data is initially presented in a chart. Any element in the chart can be selected with the mouse to show all the associated data in a separate information window. The data can be scrolled through and/or expanded in, the chart window.

Automatic title generation: The title generator allows construction of saveable title templates from any combination of constant strings and variable parameters whose values are determined by the data present in the chart window. The title generator algorithm checks the data for each variable parameter in the template and looks for a unique title. If none is found a list of possible titles is presented and the user then has options for how to use the information *e.g.* concatenation can be used. There is an 'aliasing' system to allow substitution of variable data values with an alternative if the DB value is unsuitable in a title. There is also an option to condense Probe Set information for readability *e.g.* Set 1, Set 2, Set 4, Set 5, Set 6 would become Sets 1,2,4 – 6. The title generator operates automatically whenever the contents of the chart window changes, using the current template.

Sorting and filtering: The RO has efficient methods for sorting, filtering and bookmarking which the software uses through the user-interface to allow complex nested sorting and comprehensive filtering with any combination of AND and OR on built-in parameters. There are also special fields returned by the RO used to construct fully customized sorters and filters. The lists of built-in sorters and filters are shown in Figure 5.

Panels: To highlight divisions of data, coloured panels can be added behind the histogram elements of charts and legends may be added. There is a drop-down list of parameters to choose from and colour schemes can be constructed and saved. The panel generator operates by building a bitmap behind the chart elements, using the current colour scheme, after analyzing the chart data with respect to the parameter to be panelled. For example, if a fucose panel were used the panel colour for a particular histogram element would be set by the chosen colour scheme based on the number of fucose monosaccharides present on the probe. In Figure 7 sialyl-linkage panels are used to show divisions between groups of glycan probes containing, 2 – 3, 2 – 6, 2 – 3 / 2 – 6, 2 – 8 and 2 – 9 sialyl linkages with the data sorted by a sialyl-linkage sorter.

Other options: There are choices to present averages of either or both the 7 (or 5) or 2 fmol data and the presentation of both probe (x) and fluorescence (y) axes. For example, the amplitude of the fluorescence axis can be chosen to have an absolute value or one related to the peak maximum in the chart data and the numbering of the probe axis can be selected to reflect position in the chart, position in the total retrieved data, position before applying sorters and filters or the relation to a master arrangement defined by a custom sorter. Filtered or unfiltered data can be excluded or shown highlighted for differentiation.

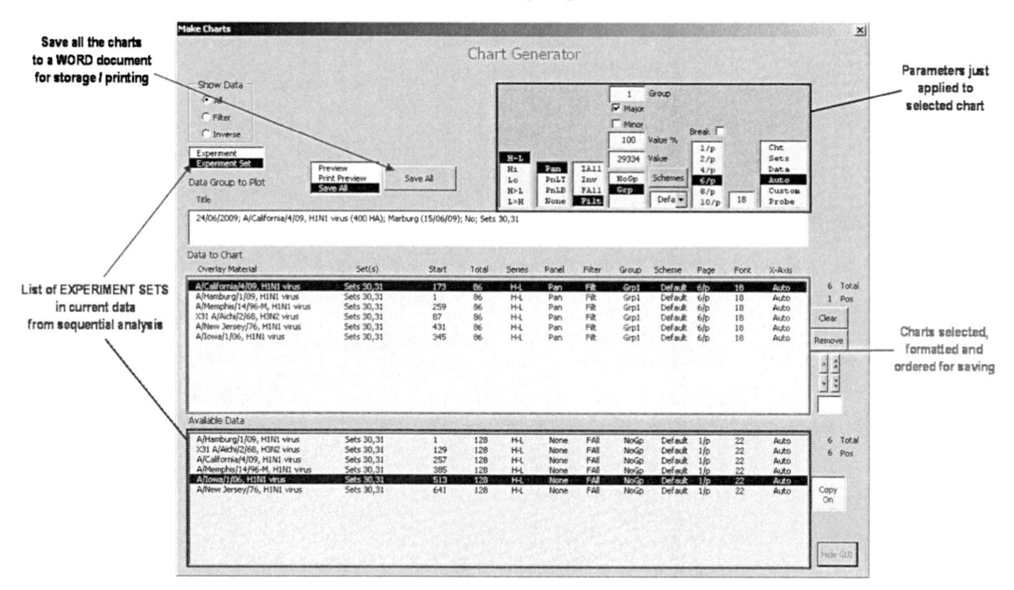

Figure 6. The user-interface of the chart generator. Data are selected from 'Experiment' or 'Experiment Set'. The items under those headings show in the 'Available Data' window. Items of each type can be transferred in the required order to the 'Data to Chart' window where each item is formatted using the parameters shown in the top right of the interface. There is control over binding value presentation, data panels, filter presentation, y-axis values and grouping, chart colour schemes, number of charts per page, title font size and x-axis format. The chosen charts can be previewed in the chart window, printed directly or saved to a WORD document (Figure 7).

Automatic chart generation: There is a comprehensive interface to set-up automatic chart generation from the current data and sorter/filter configuration (Figure 6). There is a choice to divide data into 'Experiments' and/or 'Experiment Sets' and to apply individual presentational parameters to each chart chosen for saving or printing. All titles are generated using the current template but are editable before outputting. Charts are saved in WORD (Figure 7) as in-line enhanced metafiles. The size of each chart is determined by the charts per page parameter applied to each chart before saving. The chart generator algorithm gathers all the parameters to be applied to make each chart in sequence and after hiding the chart window from view generates each chart in turn, converts a copy of it to a metafile and adds that into a new, formatted WORD document, which has been created by the software for reception of these objects. The SAVE AS dialogue is then presented to the user.

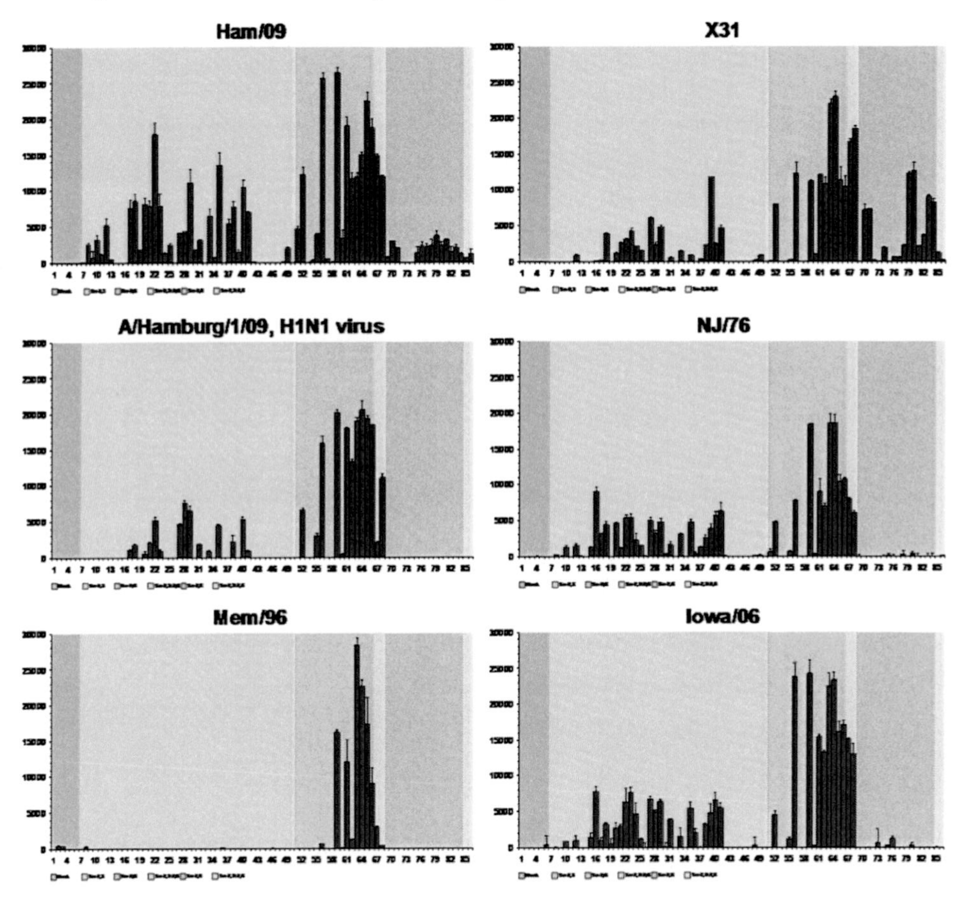

Figure 7. Six charts on a page produced by the chart generator with the settings shown in Figure 6. A sialyl-linkage sorter was used to group the glycan probes. The y-axis shows the fluorescence intensity of the binding signals from 6 influenza viruses. The legend shows the panel colours for non-sialylated and $\alpha2-3$, $\alpha2-6$, $\alpha2-3/\alpha2-6$, $\alpha2-8$ and $\alpha2-3/\alpha2-8$ sialyl-linked probes. The titles are generated using a protein or virus name template with aliasing to give the wording shown. (taken from [12])

Tables: Tables can be generated from any concurrent region of the data using any selection and order of parameters from a list provided. Repetitions of parameters are allowed. Titles for columns are automatically supplied and the tables are available for immediate printing or copying and pasting to suitable containers. A useful feature is that a master table can be constructed which can be referenced numerically from other data presentations so the table can be used to determine information for any data point in a chart, matrix or other table which may be a rearranged subset of the master table (Figures 8 and 9).

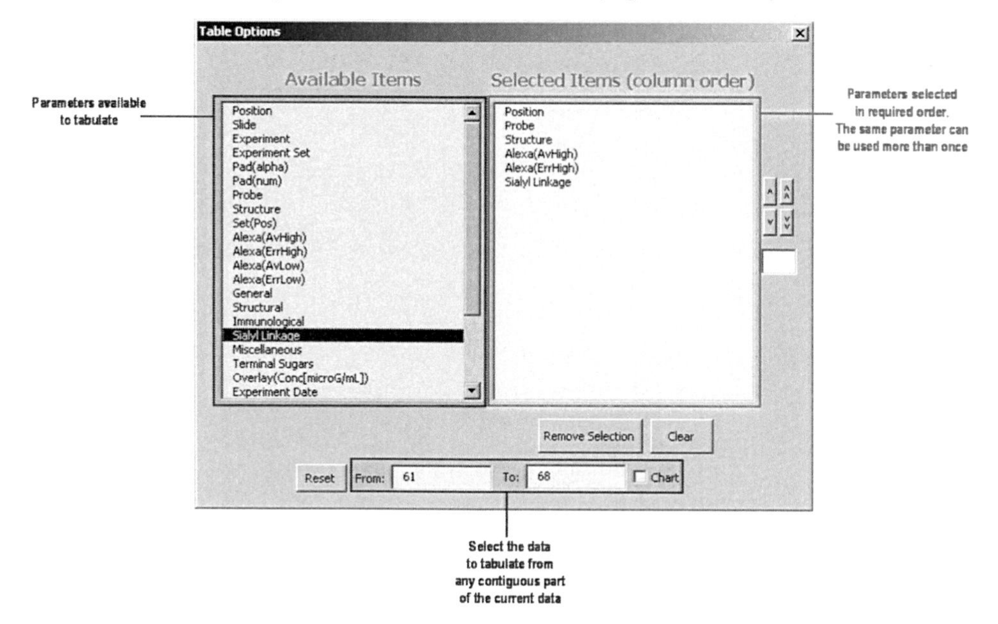

Figure 8. The user-interface of the table generator. 'Available Items' are selected from the list in any order. The data to tabulate can be that present in the chart window or any contiguous region of the current data. Column headings, as in Figure 9, are automatically generated.

Position	Probe	Structure	Alexa (AvHigh)	Alexa (ErrHigh)	Sialyl Linkage
61	SA(6')-LNFP-VI	NeuAcα-6Galβ-4GlcNAcβ-3Galβ-4Glc Fucα-3	17951	100	2,6-Sialyl termini
62	MSLNH	NeuAcα-6Galβ-4GlcNAcβ-6 Galβ-4Glc Galβ-3GlcNacβ-3	13303.5	310.5	2,6-Sialyl termini
63	MSLNnH-I	Galβ-4GlcNAcβ-6 Galβ-4Glc NeuAcα-6Galβ-3GlcNacβ-3	18948	568	2,6-Sialyl termini
64	DSLNnH	NeuAcα-6Galβ-4GlcNAcβ-6 Galβ-4Glc NeuAcα-6Galβ-3GlcNacβ-3	20490	1339	2,6-Sialyl termini
65	MFMSLNnH	Galβ-4GlcNAcβ-6 Fucα-3 Galβ-4Glc NeuAcα-6Galβ-3GlcNacβ-3	19284	510	2,6-Sialyl termini
66	A2(2-6)	NeuAcα-6Galβ-4GlcNAcβ-2Manα-6 Manβ-4GlcNacβ-4GlcNac NeuAcα-6Galβ-4GlcNacβ-2Manα3	18423.5	13.5	2,6-Sialyl termini
67	DST	NeuAcα-3Galβ-3GalNAc NeuAcα-6	2076	130	2,3:2,6-Sialyl termini
68	A3	NeuAcα-3Galβ-4GlcNAcβ-2Manα-6 Manβ-4GlcNacβ-4GlcNac NeuAcα-3Galβ-4GlcNacβ-4Manα3 NeuAcα-6Galβ-4GlcNAcβ-2	11033.5	526.5	2,3:2,6-Sialyl termini

Figure 9. A table showing data between data points 61 and 68 inclusive of the current data (Figure 8). The items included are: position in the chart, glycan probe name, probe structure, average binding signal at 5 fmol, error in the 5 fmol duplicates and the 'Sialyl Linkage'.

Matrices: A matrix is defined as a 2D array plot with proteins along one axis and glycan probes along the other (Figures 10, 11). The unique feature is that probes of the same structure from different probe sets are lined up so that binding intensities for a given glycan structure can be easily visualized for each protein in turn. To achieve this, blank spaces are inserted where corresponding data are not available. Matrices are generated in an EXCEL worksheet and are fully interactive so that all data associated with any data point can be viewed in the information window. There are a variety of options including custom colour schemes, matrix- or protein-wide normalization, inclusion of extra parameters such as master table numbers and glycan structures and highlighting of specific data as desired using custom filters.

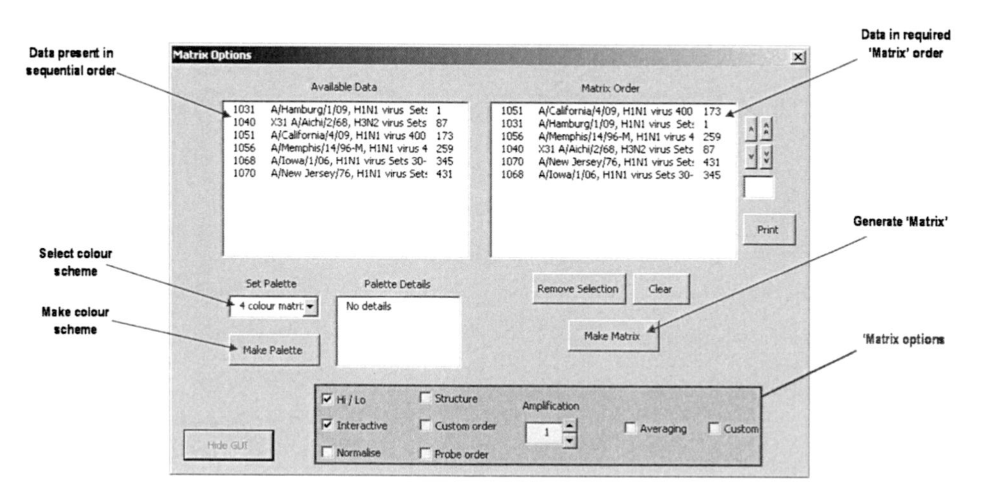

Figure 10. The user-interface of the 'matrix' generator. 'Available Data' are listed and items are transferred to and rearranged as required in 'Matrix Order. A colour scheme is made or selected from those saved and options are selected. There is control of: data to use (5 or 2 fmol), normalization over all or each protein or using selected data points, highlighting of custom items, inclusion of probe structures and special reference position information. The matrix may be interactive if desired to show all data associated with any selected data point. The data are presented in an EXCEL worksheet as in Figure 11.

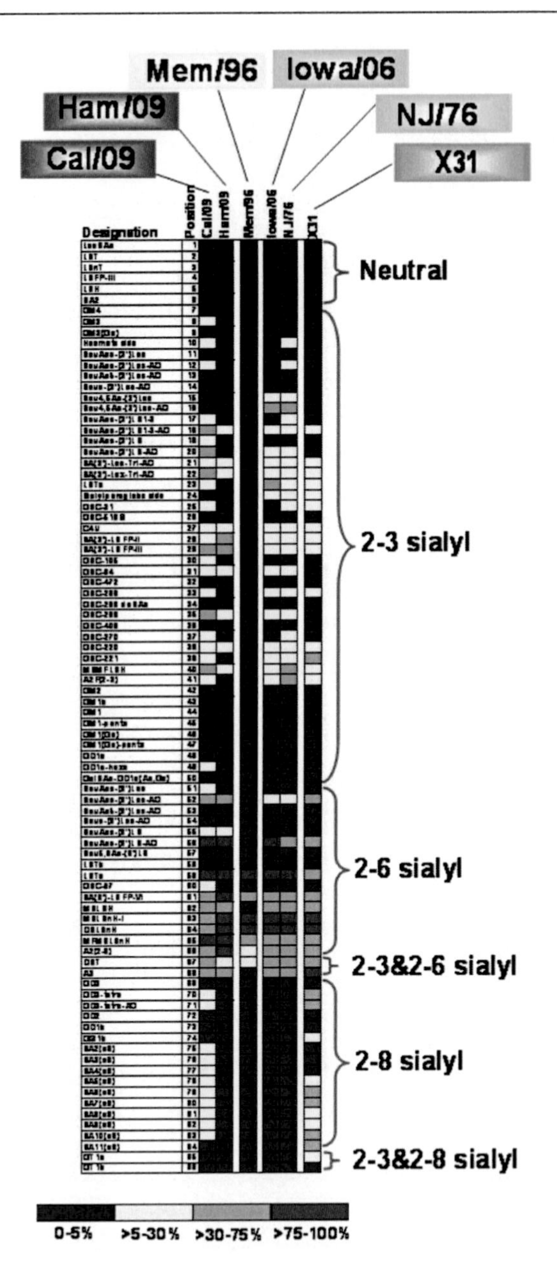

Figure 11. Matrix generated with the settings shown in Figure 10. It shows the same data as in Figure 8 but in a more compact form and more importantly could if necessary apply to data derived from different probe sets which would be more difficult to interpret from charts alone. The relative binding intensities in the matrix were calculated as the percentage of the fluorescence signal intensity at 5 fmol given by the probe most strongly bound by each virus. (taken from [12]).

Slide graphics: There is an option to view all slide data as a slide graphic similar to that produced by the scanner software with a magnified experiment graphic (Figure 12) such that the intensity of each data point is shown as a colour gradient from black to red for the binding signal or black to green for the reference signal. The scale can be selected from linear, quadratic or quartic. The latter two allow enhancement of lower binding levels. There is an option to show negative values in graduated black to blue. Normalization can be slide-wide or experiment-wide. The system is fully interactive showing all the data in the DB for any selected data point. The slide and experiment graphics can be captured as bitmaps for storage or printing. A particular use of this feature is as a way to check new data entry against the original picture of the slide generated by the scanner. The two patterns should correspond. It also offers a useful alternative way to view all the raw data from a slide in a compact interactive format.

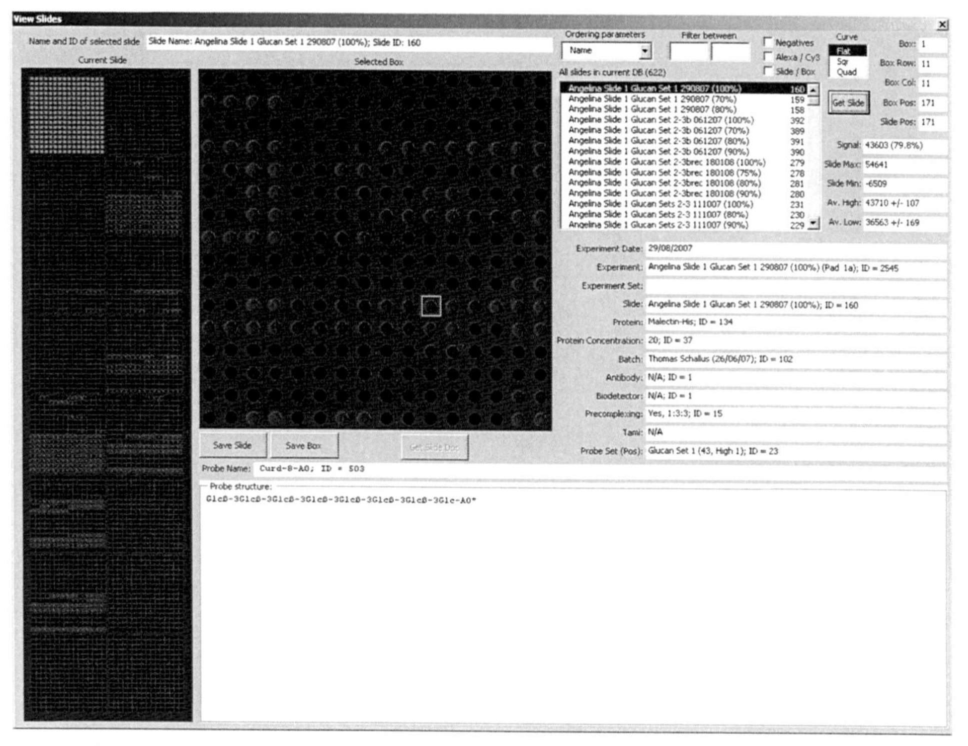

Figure 12. The slide graphic data presentation. The list shows every slide in the database. Selecting a slide will show all its data at once in the same form as that produced by the laser scanner. The enlarged box shows one of the 16 experiments in the slide. Each experiment can be selected in turn. The small yellow box shows the microarray spot selected and all the data associated with that spot are shown on the right hand side of the interface. In the graphic, fluorescence values are shown as black (no signal) to red (maximum binding signal) or green (maximum reference signal). There are options to show the gradient as linear, quadratic or quartic to enhance low values, to show negative values in blue and to normalize over the whole slide or each experiment. The graphics images can be saved as bitmaps for storage.

Conclusions

The software suite described here is an indispensable tool in support of our carbohydrate microarray programme. It allows rapid analysis of new data and comparison with all stored information. The system at present is tailored to our specific needs and is designed around our screening format.

General purpose multi-format software

We are currently developing a new software suite aimed at much more general carbohydrate microarray systems that can handle a wide range of microarray layouts and that will be able to analyse dose-response and inhibition data in any format. We are collaborating internationally with scientists who are experts in bioinformatics in an endeavour to make this new software available to the glyco-community.

Acknowledgments

We are grateful to our colleagues for continual feedback through usage which has led to many improvements and valuable additions to the software during development. In this regard we acknowledge, Maria A. Campanero-Rhodes, Angelina S. Palma, Robert A. Childs, Yan Liu, Wengang Chai and Alexander M. Lawson. This work was supported by grants from UK Research Councils' Basic Technology initiative 'Glycoarrays' (GR/S79268); UK Engineering and Physical Research Councils Translational Grant EP/G037604/1; UK Biotechnology and Biological Sciences Research Council (BB/E02520X/1 and BB/G000735/1); UK Medical Research Council (G9601454 and G0600512); Wellcome Trust (WT085572MF, NCI Alliance of Glycobiologists for Detection of Cancer and Cancer Risk (U01 CA128416) and Fundação para a Ciência e Tecnologia, Portugal (SFRH/BPD/26515/2006)

References

[1] Tang, P.W., Gooi, H.C., Hardy, M., Lee, Y. C. & Feizi, T. (1985) Novel approach to the study of the antigenicities and receptor functions of carbohydrate chains of glycoproteins. *Biochem. Biophys. Res. Commun.* **132**:474 – 480.
doi: http://dx.doi.org/10.1016/0006-291X(85)91158-1.

[2] Feizi, T., Stoll, M.S., Yuen, C.-T., Chai, W. & Lawson, A.M. (1994) Neoglycolipids: probes of oligosaccharide structure, antigenicity and function. *Methods Enzymol.* **230**:484 – 519.
doi: http://dx.doi.org/10.1016/0076-6879(94)30030-5.

[3] Stoll, M.S., Feizi, T., Loveless, R.W., Chai, W., Lawson, A.M. & Yuen, C.T. (2000) Fluorescent neoglycolipids. Improved probes for oligosaccharide ligand discovery. *Eur. J. Biochem.* **267**:1795 – 1804.
doi: http://dx.doi.org/10.1046/j.1432-1327.2000.01178.x.

[4] Chai, W., Stoll, M.S., Galustian, C., Lawson, A.M. & Feizi, T. (2003) Neoglycolipid technology – deciphering information content of glycome. *Methods Enzymol.* **362**:160 – 195.
doi: http://dx.doi.org/10.1016/S0076-6879(03)01012-7.

[5] Feizi, T. & Chai, W. (2004) Oligosaccharide microarrays to decipher the glyco code. *Nat. Rev. Mol. Cell Biol.* **5**:582 – 588.
doi: http://dx.doi.org/10.1038/nrm1428.

[6] Paulson, J.C., Blixt, O. & Collins, B.E. (2006) Sweet spots in functional glycomics. *Nat. Chem. Biol.* **2**:238 – 248.
doi: http://dx.doi.org/10.1038/nchembio785.

[7] Horlacher, T. & Seeberger, P.H. (2008) Carbohydrate arrays as tools for research and diagnostics. *Chemical Society Reviews* **37**:1414 – 1422.
doi: http://dx.doi.org/10.1039/b708016f.

[8] Liu, Y., Palma, A.S. & Feizi, T. (2009) Carbohydrate microarrays: key developments in glycobiology. *Biol. Chem.* **390**:647 – 656.
doi: http://dx.doi.org/10.1515/BC.2009.071.

[9] Goldberg, D., Sutton-Smith, M., Paulson, J. & Dell, A. (2005) Automatic annotation of matrix-assisted laser desorption/ionization N-glycan spectra. *Proteomics* **5**:865 – 875.
doi: http://dx.doi.org/10.1002/pmic.200401071.

[10] Maass, K., Ranzinger, R., Geyer, H., von der Lieth, C.W. & Geyer, R. (2007) "Glyco-peakfinder" – *de novo* composition analysis of glycoconjugates. *Proteomics* **7**:4435 – 4444.
doi: http://dx.doi.org/10.1002/pmic.200700253.

[11] Aoki-Kinoshita, K.F. (2008) An introduction to bioinformatics for glycomics research. *PLoS. Comput. Biol.* **4**:e1000075.
doi: http://dx.doi.org/10.1371/journal.pcbi.1000075.

[12] Childs, R.A., Palma, A.S., Wharton, S., Matrosovich, T., Liu, Y., Chai, W., Campanero-Rhodes, M.A., Zhang, Y., Eickmann, M., Kiso, M., Hay, A., Matrosovich, M. & Feizi, T. (2009) Receptor-binding specificity of pandemic influenza A (H1N1) 2009 virus determined by carbohydrate microarray. *Nat. Biotechnol.* **27**:797 – 799.
doi: http://dx.doi.org/10.1038/nbt0909-797.

Glyco-Bioinformatics – *Bits 'n' Bytes of Sugars*
October 4th – 8th, 2009, Potsdam, Germany

SysBioWare: Structure Assignment Tool for Automated Glycomics

Sergey Y. Vakhrushev[1,*], **Denis Dadimov**[2], **Jasna Peter-Katalinić**[3,4,#]

[1]EMBL, Genome Biology Unit, Quantitative Proteomics, Meyerhofstrasse 1, 69117 Heidelberg, Germany

[2]MechSystemProject U.E., Bioinformatics division, 160 – 56 Mayakovsky str, Minsk 220028, Belarus

[3]University of Rijeka, Department of Biotechnology, S. Krautzeka bb, 51000 Rijeka, Croatia

[4]University of Münster, Institute of Pharmaceutical Biology and Phytochemistry, Hittorfstr. 56, 48149 Münster, Germany

E-Mail: [*]sergey.vakhrushev@embl-heidelberg.de, [#]jkp@uni-muenster.de

Received: 26th April 2010 / Published: 10th December 2010

Abstract

Glycomics as a part of systems biology closely related to proteomics encompasses knowledge acquainted by comprehensive and systematic studies of entire complement of carbohydrates in a cell, organ or organism. The prerequisite for these studies is a detailed information on molecular structure of complex carbohydrates which play a crucial role in processes like signalling, cell-cell recognition and immune response, and which act as therapeutic agents, vaccine or drug targets. Although the modern methods of mass spectrometry are well fitted for integrative "omics" experiment design, the interpretation of carbohydrate mass spectral data is still strongly linked to the human expertise. In this chapter we present a computational approach for automatic interpretation of mass spectral data of complex carbohydrates. We contribute to the field by designing a software package which will significantly reduce a need for human expertise in mass spectrometric data interpretation derived from glycoconjugates and enable discovery and improvement of high-throughput protocol for automated glycomics. The proposed structure assignment tool named SysBioWare was constructed for automated processing raw MS and MS/MS performing isotopic grouping of detected peaks after de-noising and wavelet analysis. Monoisotopic *m/z* values render peak list association with the raw MS spectrum and allow compositional assignment according to the tuned building block library. This platform has been applied to human urinome and glycolipidome as a potent tool for rapid assignment of already known or/and new carbohydrate structures.

Introduction

Oligosaccharides, glycolipids, glycoproteins, glycopeptides, and proteoglycanes (mucopolysaccharides) occur ubiquitously in nature. Most of the naturally occurring proteins are glycosylated and these glycoproteins are found together with glycolipids in viruses, microorganisms, plants, and animals. Whereas the cell surface of bacteria, yeast, plants and other lower forms of life is primarily composed of polysaccharides, thus forming rigid cell walls, the cell membrane of mammalian cells contains carbohydrate residues linked to proteins and lipids. The ceramide residue (i. e., a long chain sphingoid base substituted at the amino group by a fatty acid) is by far the most prominent lipid constituent in addition to minor quantities of glycerol ethers and esters. In glycoproteins, glycan chains can be linked either N-glycosidically to asparagine residues or O-glycosidically to serine or threonine.

The growing perception of the biological importance of this group of compounds has greatly stimulated efforts in developing new and more powerful methods of structural elucidation. However, compared to the progress achieved in the structural analysis of other biopolymers such as proteins or nucleic acids, progress has been rather slow in the field of complex carbohydrates. This is mainly due to the complexity of carbohydrate structures, a consequence of the polyfunctional sugar molecule that allows several sites of anomeric linkages.

A complete structural analysis of a complex carbohydrate molecule involves the determination of (a) molecular mass and number of individual sugar components, (b) sites and anomeric configuration of glycosidic linkages, (c) conformation of sugar rings, (d) sequence of sugar components and pattern of branching, (e) secondary structure and spatial orientation, and (f) structure of the aglycon. Among methods for glycoanalysis, mass spectrometry have been found to be especially powerful for determining (a) molecular masses (b) sequence and pattern of branching, (c) in special cases, sites of glycosidic linkages and (d) structure of the aglycon. From the molecular mass, the number of sugar constituents in terms of deoxy hexose, hexose, hexosamine, etc. can, in most cases, be calculated [1]. Presently, MS is a most popular method for structural analysis due to its high sensitivity and speed on the one hand, and the ability to analyze complex mixtures on the other. In particular, qualitative data interpretation of MS spectra in high-throughput projects appears of primary importance for rapid identification of biological routes [2].

Glycomics encompasses comprehensive and systematic genetic, physiological, pathological and structural studies of entire complement of carbohydrates in a cell, organ or organism. Although the modern MS instruments are fitted for integrative "omics" experiment design, the interpretation of carbohydrate mass spectral data is still strongly linked to the human expertise, and due to the carbohydrate structure complexity, it represents a complex task for specialists. In order to extent the limits and options of systems biology, efficient tools for glycomics are required. In this chapter we present a computational approach for automatic interpretation of mass spectra of complex carbohydrates to be used in glycomics.

Availability of Tools for Interpretation and Deposition of Complex Carbohydrate Structure

Over the past 20 years several groups reported on their attempts to develop rational tools for computational mass spectrometry in glycan analysis. Various products focus on compositional analysis, structure drawing, data base development, *in silico* fragmentation and spectra assignment, but very limited number of them efforts to incorporate rational tools into a single package. In some recent reviews the state-of-the-art in glycoinformatics, particularly in the field of computation of MS data, has been summarized [2 – 5].

In the "Glycomod" tool all possible glycan compositions are calculated from the values of their respective molecular ions, which can be done for the native permethylated and peracetylated structures. For glycopeptides, "Glycomod" infers compositional candidates if the mass and/or the sequence of the parent peptide is known [6, 7]. Upon the automatic creation of the composition hit list, it must be further manually analysed to remove from the hit list implausible glycan portion structure proposals, since a step for filtering of biologically non-relevant carbohydrate compositions is not provided.

"GlycoPep DB" has been designed as a web-based tool for glycopeptide analysis using a "smart search" concept where the human expertise for filtering implausible structures should be largely reduced. Designed for N-glycopeptide compositional assignment, this program can compare experimentally determined masses against the database of glycopeptides with N-linked glycans, where only biologically relevant structures are stored [8]. In this way the number of implausible glycan compositions in "GlycoPep DB" in comparison to "Glycomod" is reduced, but this approach is functionally limited to those structures, which are already known, and therefore present in databases.

Using "GlycoSuite DB" the glycan composition can be assigned to the experimentally determined selected precursor ion, which is present in the glycan structure database either on the basis of the mass search through glycan structure database or is deposited according to its composition. According to the last update in May 2009, in this database 9436 entries from 864 published references were collected. Additionally, the query for the particular glycan can be performed either by the sequence, or the type of glycosidic linkage, or its biological source, accession, and taxonomy, by the type of disease, the attached protein and is associated with the appropriate reference (http://glycosuitedb.expasy.org/glycosuite/glycodb). For interpretation of oligosaccharide fragmentation patterns the "GlycosidIQ^{TM}" software, based on the principle of matching experimental MS/MS data with those generated *in silico* from the GlycoSuite DB, is available [9]. The output is ranked according to proprietary algorithm which takes into account the fragmentation properties of branched glycans as well as their biological probabilities.

Based on a similar approach "GlycoFragment" is another tool by which the generation of all theoretically possible A-, B-, C-, X-, Y- and Z-fragments of oligosaccharides can be performed and each peak of a measured mass spectrum compared with the calculated fragments of all structures contained in the SweetDB database [10, 11].

In "STAT" a list of all possible saccharide compositions based on its mass alone will be generated. To find all possible N-linked oligosaccharide structures for this particular composition, the fragments imputed manually can be analysed by rejecting all structures not containing a trimannosyl core. However, other biosynthetic pathway rules are not included. Fucosylated glycans and glycans containing a bisecting GlcNAc residue are not considered [12].

The web-based application "Glyco-Peakfinder" was developed for a rapid assignment of glycan compositions. To provide entirely *de novo* compositional assignment, this platform does not operate with prior information from glycan databases or pre-calculated known archetypes. In the computational stage it can accept a number of glycan derivatizations such as permethylation, peracetylation, perdeuteromethylation, and acetylation [13]. This tool is complemented by "GlycoWorkbench", which is a suite of software tools designed for rapid drawing of glycan structures and for assisting the process of structure determination from mass spectrometry data. The graphical interface of GlycoWorkbench provides an environment in which structure models can be assembled. Mass values from the structure candidate list can be computed, their fragments automatically matched with MS^n data and the results compared to assess the best candidate followed by MS/MS data annotation.

Another development based on matching of *in silico* generated fragments against measured MS/MS data is a commercial product "SymGlycan". Candidate structures for *in silico* fragmentation are retrieved from the proprietary database of biologically plausible structures.

Each candidate structure is scored to indicate how closely it matches the experimental data. Additionally, this program provides elements of project management, for associating results with input profile and managing search parameters [14].

As long as updates of carbohydrate databases are still far behind developments in proteomics and because *in silico* approach imposes functional limitations on application to glycomics unlike in proteomics, computational mass spectrometry for glycomics is highly focused on the development of tools for *de novo* glycan MS/MS interpretation.

"StrOligo" is a tool for automated interpretation of tandem MS spectra of complex N-linked glycans. Tandem mass spectra are simplified leaving only monoisotopic peaks and based on assumption that all glycans contain a trimannosyl core. The remaining number of GlcNAc in the composition will determine the number of antennae, where the possibility of bisecting GlcNAc is eliminated if the number of remaining Gal is the same as the number of GlcNAc. If the number of Gal or NeuAc or NeuGc is lower than the number of antennae, the different structural isomers are assessed. The introduction of polylactosamine repeats and branch fucosylation is allowed and experimental mass spectra will be evaluated by matching them with those generated *in silico* [15, 16].

Another program for automated annotation of glycan MS data is "Cartoonist", which was developed for the automated annotation of N-glycan MALDI TOF mass spectra. "Cartoonist" is constructed on the principle of labelling MALDI peaks with cartoons representing the most plausible glycan assemblies biosynthesized in mammals using 300 manually determined archetypes [17]. By this approach the numbers of implausible structural candidates for a certain precursor ion is largely reduced, but on the other hand due to the fixed library of created archetypes, its application area is limited by the size of the library itself. Automated identification of N-glycopeptides was realized by extending Cartoonist to another program called "Peptoonist" using a combination of MS and MS/MS data [18].

All previously mentioned tools are mostly focused on single functional task like compositional proposal, data base search and spectra annotation. Moreover, laborious spectrum processing tasks such as peak recognition and charge deconvolution still require manual user intervention in most cases.

The SysBioWare suite presents first platform which pipelines all crucial steps of computational mass spectrometry into one process, in particular:

- raw spectrum processing (baseline adjustment and noise removal),
- peak shape detection using continuous wavelet analysis,
- charge deconvolution and isotopic clustering,
- proposing candidate compositions,

- filtering biologically implausible candidates,
- generating candidate structures using *de novo* approach or open database search,
- assessing candidate structures using *in silico* fragmentation,
- providing reporting facilities for comparing composition lists, etc.

SysBioWare is accessible in a web-based interface or as a desktop application. Desktop version additionally provides data storage and management modules which allows user to organize experiments in tree-like structures with annotation and searching facilities. Also, user can tune analysis parameters to adjust to specific hardware and sample and save parameter sets as analysis profiles. Calculation-intensive tasks such as shape detection and isotopic distribution calculation are implemented using Intel Performance Primitives library which fully exploits processor hardware optimizations [19].

Here is a typical analysis scenario in SysBioWare:

1. User loads MS spectrum data. mzXML or plain ASCII formats are currently supported.
2. User chooses peak detection profile that best suits hardware used and experiment settings.
3. Next, user can run peak detection as a single step. In our experiments this step typically completed within one minute providing true positive rate in the range of 85 – 95%. Additionally, user can step through individual processing stages and adjust individual parameters. This is useful for developing and debugging analysis profiles.
4. User chooses compositional analysis profile that best describes sample substance used in the experiment. This profile determines which building blocks (monosaccharides, adducts and modifications) will be tried and describes plausible biological structures as a set of rules. Every profile element can be adjusted if necessary.
5. Candidate composition list is automatically produced. Since it is based on profile selected at the previous step, only feasible compositions are retained. In our experiments, we achieved one or two compositions per assigned peak.
6. If the goal of this experiment is structure identification, user selects precursor ion and most likely composition and proceeds to MS/MS analysis phase.
7. Here, as in MS-analysis step, user loads MS/MS data and runs peak detection module first.

8. Then, user selects MS/MS analysis profile. This profile sets *de novo* synthesis and *in silico* fragmentation parameters. Alternatively, user can construct molecule structure interactively or block by block using IUPAC notation
9. *De novo* structure generation algorithm then generates candidate structure list which is ranked to bring up the structures which have *in silico* fragmentation pattern closest to observed MS/MS data.
10. The top structure is selected and passed on to *in silico* fragmentation and annotation phase. Here structure and spectrum can be annotated with individual fragments.

The SysBioWare web site www.sysbioware.com provides access to all functionality described. It also provides video introduction and demo analysis sessions for selected carbohydrates structures.

Architecture and General Features of SysBioWare Demonstrated on N-Glycan Mixtures

Computational Mass Spectrometry for Glycomics

The software utilities are constructed of distinct modules dealing with mass spectrometric data processing and their interpretation (Figure 1). *"Composition Blocks"* constructor (Figure 1) allows users to create a library of potential building block components, adducts and modifications, which can be used at further analysis steps. Component increment masses are calculated automatically from elemental composition formula. *"Molecule Classes"* library defines a list of molecule classes to be observed (Figure 1, inset). It is defined by: a) the list of all possible building blocks, which provides the foundation for compositional assignment; b) biological feasibility rules (Bio-filter). These rules, written as a line code, specify conditions or the ratio between building blocks; c) an average elemental composition model is used for more accurate isotopic peak grouping. At the current stage two types of model biomolecules are implemented: peptides and glycans. *"Laboratory Module"* provides user interface to the database of experiments.

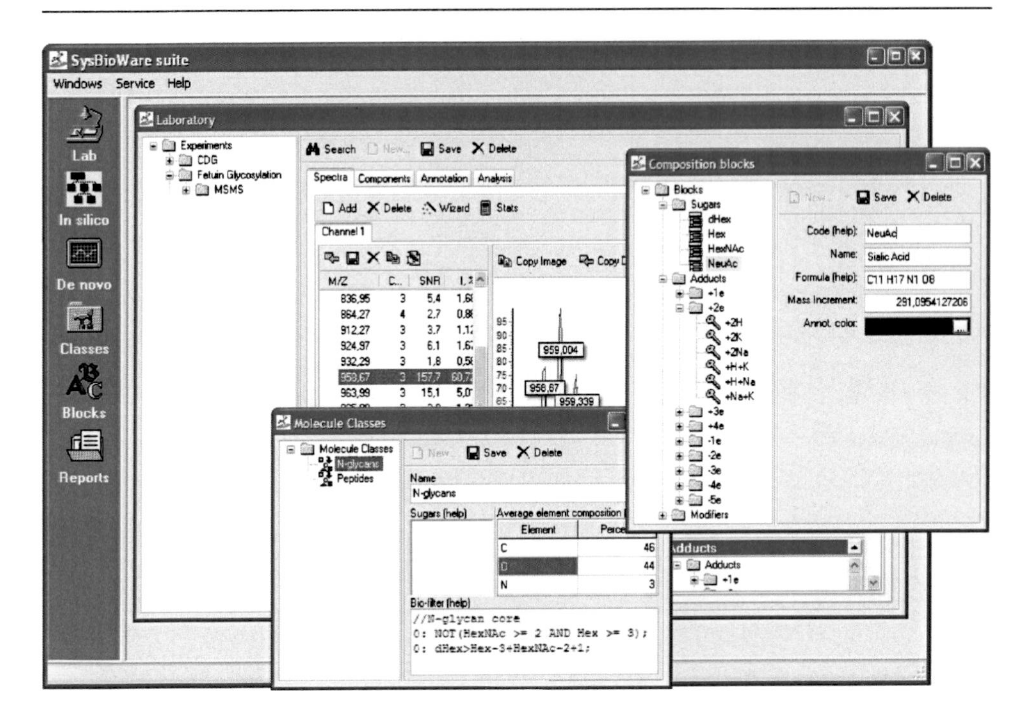

Figure 1. The main window of the SysBioWare 2.0.0 software with activated "Library", Compositions blocks" and "Molecule Classes" modules.

The user starts by importing raw mass spectrum at the "Spectra" tab, where Peak Detection Wizard generates initial monoisotopic peak list. The "Components" tab requires users to provide the lists of building blocks, adducts and modifications they are expecting to see as well as the list of bio-filter rules which can be imported from the molecule classes library. "Annotation" tab allows the user to describe various aspects of the experiment such as materials, methods, instrument settings, biological sources, etc. These annotation elements can be later searched to retrieve past experiments and make cross-references. Finally, "Analysis" tab is the place where compositional assignment is performed. This procedure is based on the principle of modelling of the respective glycoconjugate ions using different combinations of potential building blocks defined by the user in the "Components". Afterwards, assignment results can be compared between different experiments and exported to Excel.

Peak Detection Wizard

The wizard provides mass spectrometric data processing (base line correction and noise level determination), peak shape determination (using continuous wavelet transform), automated monoisotopic *m/z* values recognition and charge state determination (Figure 2).

Data import and resampling. The raw spectrum needs to be cleaned from duplicates *m/z* values (spectrum points with the equal *m/z* values and non-equal intensity values) and re-sampled to have uniform *m/z* values. A number of optimization steps, such as the speed of wavelet analysis calculation, are enhanced (Figure 2A-B).

Baseline Correction. We have introduced two modes of baseline correction, which can be optionally selected by the user. The first mode detects the electronic noise, which is assumed to have normal distribution and then uses this noise as a lowest signal level. The second mode is based on the removal of specified proportion of lowest measurements. Afterwards, the baseline can be subtracted from the spectrum (Figure 2C).

Smoothing and Peak Detection. At this step, noise spikes and coarse quantization effects are removed from the spectrum (Figure 2D).

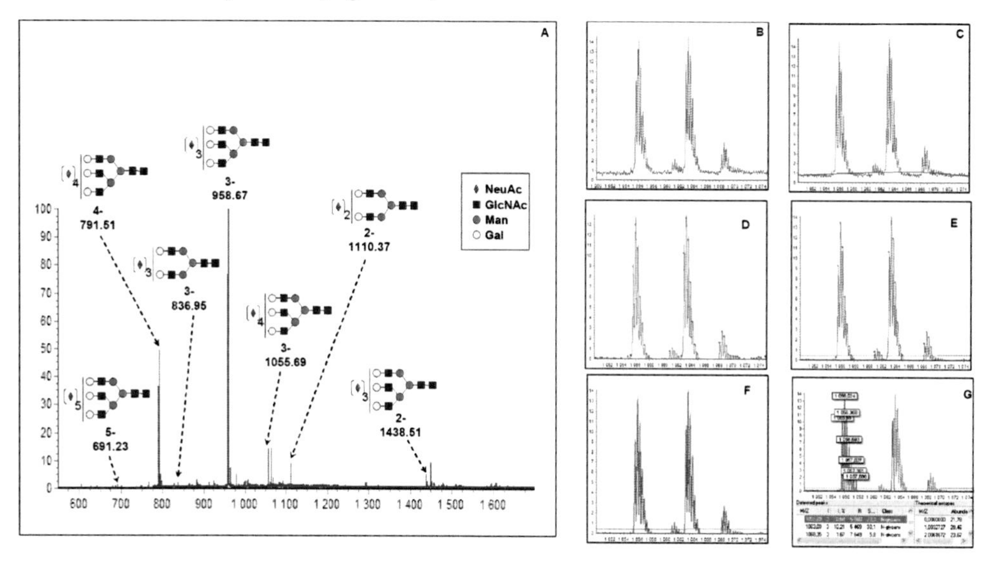

Figure 2. (A) The (-) nanoESI Q-TOF MS data of the mixture of N-glycan digested from the bovine fetuin by PNGase F. Different stages of the working process of Peak Detection Wizard are shown on the example from zoomed area of the raw spectrum (Figure 2A) in the *m/z* range from 1050 to 1075. (B) Data import and resampling stage; (C) Baseline Correction stage; (D) Smoothing and peak detection stage (shown in green); (E) Noise level determination. Peaks detected above the noise level are shown by red circles; (F) Peak Shape Detection stage based on the continuous wavelet analysis. Signals matched by Gaussians are shown in red; (G) Isotopes grouping stage: grouping of different peaks into a single isotopic envelope and charge state determination. The raw spectrum after baseline subtraction and smoothing is shown in blue. Signals matched by Gaussians are shown in grey. The recognized monoisotopic peaks are shown in red.

Noise level determination. The intensity of noisy peaks is assumed to follow normal distribution with varying local standard deviation. The mean of the noise is estimated as the median of peak intensities, and local standard deviation is estimated from the local median absolute deviation. The peaks which intensity exceeds specified percentile are considered to be useful signals (Figure 2E).

Peak detection. Peaks that have passed noise filtering are matched against one of the expected peak shapes (Figure 2F). Currently, Gaussian and Lorentzian shapes are used. Shape matching is implemented using continuous wavelet transform.

Isotope Grouping. After detecting potential peaks by shape matching, SysBioWare attempts to group them as isotopes and determines their charge state (Figure 2G). Starting with a maximum possible charge state, the program tries to find peaks at designated locations that form the packet of the same shape as a corresponding theoretical isotopic distribution. Isotopic distributions are probed by SysBioWare for different molecule classes, which can be selected by the user from the "Molecule Classes library" module, *e.g.* glycans and peptides (Figure 1). Finally, interpreted data can be referenced from subsequent experiment records and organized into table as a user report [19].

Isotopic distribution modelling

To show accuracy of approximation in oligosaccharide molecule modelling a comparison between the isotopic distributions of the real composition and computed from the model molecule has been performed (Figure 3). After baseline subtraction peak shape determination, a number of isotopic envelopes corresponding to tri-sialylated triantennary N-glycan (Figure 3A), di-sialylated biantennary N-glycan (Figure 3B) and penta-sialylated triantennary N-glycan (Figure 3C) have been compared with theoretical isotopic distribution normalised to the monoisotopic peak computed from hypothetical glycan model molecule (green dashed line). For these classes of molecules very good consistency was observed. By this, indication of the specific calculation for model molecules at masses higher than 2500 Da is demonstrated. Moreover, preliminary information about sample nature allows the user to tune the Peak Detection Wizard more precisely and correct the oligosaccharide model molecule according to any specific case.

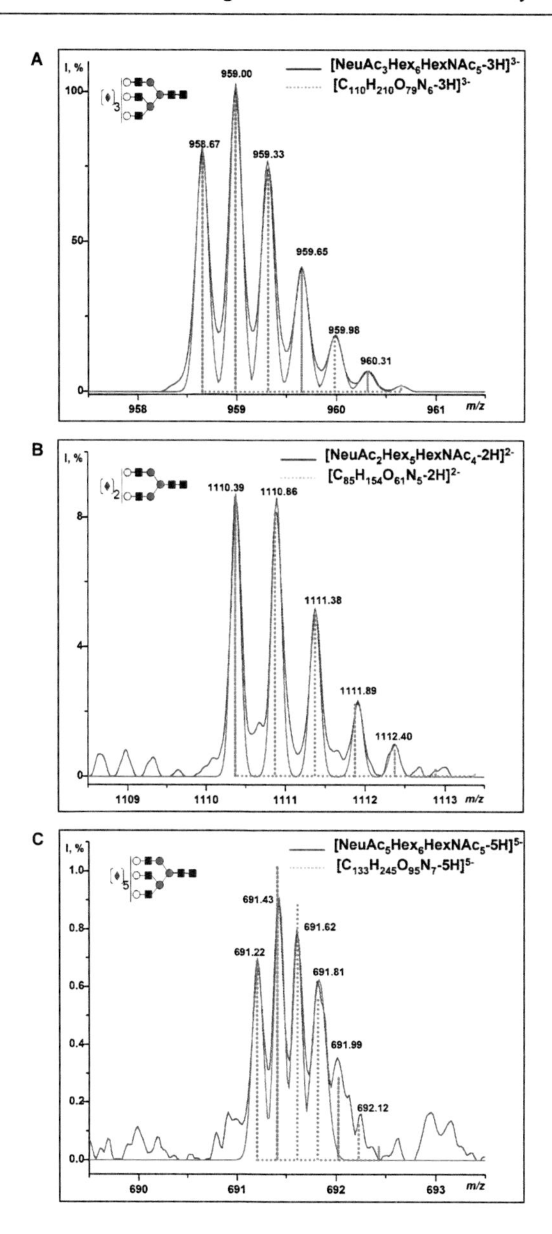

Figure 3. A comparison between baseline corrected raw spectrum from Figure 2A (blue line), peak recognised signal (red line) and isotopic distributions theoretically simulated from the hypothetical analogue calculated from the mass values based on glycan model (green dashed line). As examples the following ions have been considered: trisialylated triantennary N-glycans $NeuAc_3Hex_6HexNAc_5$ (A), pentasialylated triantennary N-glycan $NeuAc_5Hex_6HexNAc_5$ (B) and di-sialylated biantennary N-glycan $NeuAc_2Hex_5HexNAc_4$ (C). 2D glycan pictogram nomenclature is the same as in Figure 2.

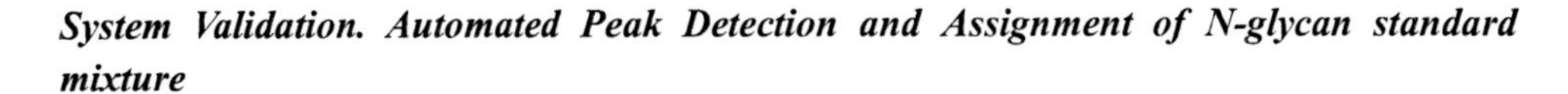

System Validation. Automated Peak Detection and Assignment of N-glycan standard mixture

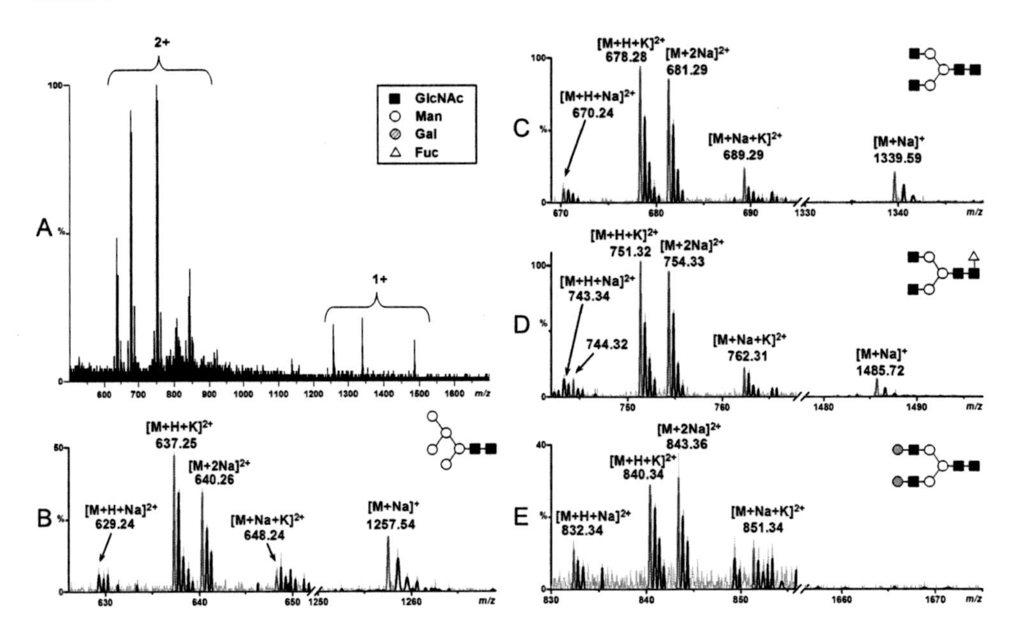

Figure 4. Example of the automated monoisotopic *m/z* peak recognition. (A) Positive ion mode nanoESI Q-Tof MS of the equimolar mixture of neutral N-glycan standards Man5, NGA2, NGA2F, and NA2. (B) Expansion of the singly (left) and doubly charged (right) detection area of the Man5 N-glycan. (C) Expansion of the singly (left) and doubly charged (right) detection area of the NGA2 N-glycan. (D) Expansion of the singly (left) and doubly charged (right) detection area of the NGA2F N-glycan. (E) Expansion of the singly (left) and doubly charged (right) detection area of the NA2 N-glycan. © 2008 American Chemical Society, reprinted with permission from [19].

To test the ability of Peak Detection Wizard in managing with high-throughput projects MS data of equimolar mixture of N-glycan standards at low mass concentration and obtained at short acquisition time (few seconds) has selected as submitted for processing. Four neutral N-glycan standards: Man5 (Catalogue No. M-00250S), NGA2 (Catalogue No. C-0720), NGA2F (Catalogue No. C-004301), NA2 (Catalogue No. C-0024300 M) purchased at (Oxford GlycoSciences, Abington, U.K.) have been dissolved each in MeOH/H_2O 1 / 1(v/v) at concentration of 0.5 pmol/µl and submitted to nanoESI Q-Tof MS analysis in the positive ion mode, simulating high-throughput procedure (Figure 4). At two scans, which correspond approximately to approximately 2 seconds of acquisition time, oligosaccharide standards have been detected as a group of singly and doubly charged ions at different signal abundances, showing the presence different forms of cations mostly formed by [M+Na], [M+K], $[M+H+Na]^{2+}$, $[M+H+K]^{2+}$, $[M+2Na]^{2+}$ and $[M+Na+K]^{2+}$ ionic species. Under the selected conditions the monoisotopic *m/z* values and charge states for all ionic species corresponding to N-glycan standards have been correctly recognized except for the $[M+H+Na]^{2+}$ at *m/z* 629 corresponding to the Man5 glycan.

Application to Glycourinomics

Turning the focus to complex carbohydrate analysis of human urine (glycourinomics) upon non-invasive sampling has been indicated by investigations, in which a large-scale accumulation of carbohydrate metabolites has been described [20]. Abnormal urinary oligosaccharide and glycopeptide excretion has been observed in clinical cases of hereditary diseases including α-*N*-acetylgalactosaminidase deficiency, G_{M1} gangliosidosis and sialidoses [21].

"Congenital Disorders of Glycosylation" is a group of inherited metabolic diseases caused by low or missing activities of enzymes involved in biosynthesis of oligosaccharides in organism [22 – 24]. Abnormal processing of carbohydrates may lead to changes in glycome profile between patient and control. Human urine was selected as a potential source for monitoring glycoconjugate pattern for the further development of methodological protocol for high-throughput screening and biomarkers discovery [25 – 31]. The one of the most important element of this strategy is application of SysBioWare computational platform for glycomics data processing and interpretation.

The validation of this platform is demonstrated on an example of complex carbohydrate analysis in urine (Figure 5). The oligomeric carbohydrate portion from the patient's urine was considered to contain possible specific metabolic components, which might be indicative for the diagnostic identification of the clinical defect. To obtain the oligomeric carbohydrate portion from the patient's urine a fractionation using gel permeation chromatography steps followed by anion-exchange chromatography has been carried out. A general characteristics of the fractions obtained was their high heterogeneity, the complexity of the profiles obtained by mass spectrometry and a high dynamic range of single components concerning the quantification [32, 33]. Usually, less than 30% of components were in the range of relative abundance 5 – 100%, where more than 70% of all components were in the range from 1 to 5%. By manual evaluation of mass spectra only more abundant components could be safely recognized according to their monoisotopic *m/z* values. Due to the presence of chemical noise clusters, the correct manual recognition in the intensity range below 1% was not reliable.

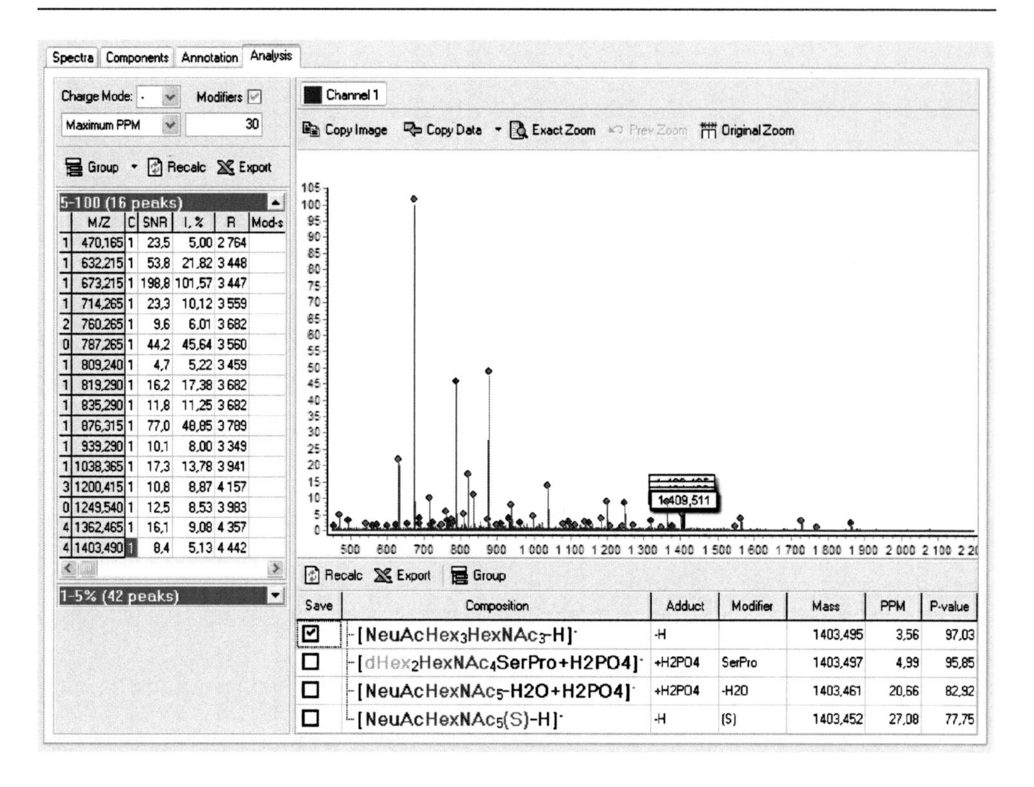

Figure 5. Example of a high-throughput screening analysis: complex glycoconjugate mixture KLM3 from the CDG patient acquired by the (-) nanoESI Q-Tof MS. Channel 1 contains data related to the KLM3 fraction. © 2008 American Chemical Society, adapted with permission from [19].

Using SysBioWare function Peak Detection Wizard the parameters were optimized to keep the detection of false positive signals as low as possible. Concerning the ions in the relative intensity range from 5 to 100% all manually determined monoisotopic peaks have been recognized correctly without any true negative and false positive signals. From peaks determined manually in the range of 1 – 5% of the relative abundance none of false positive signals have been recognized and more than 60% monoisotopic *m/z* values have been identified correctly by the Peak Detection Wizard. After the activation of the "Recalc" function within the mass deviation window less than 30 ppm, the list of all possible compositions related to the selected *m/z* values was computed. In the urine sample KLM3 from 42 molecular ions detected within the relative intensity range 1 to 5% 12 have been uniquely assigned to free oligosaccharides and 1 to a Thr-linked glycan. For seven ionic species, at *m/z* 1112.45, 1143.38, 1208.47, 1271.47, 1313.45, 1565.55 and 1727.58, respectively, two different compositions each were proposed by the platform. With respect to ionic species at *m/z* 1143.38, 1565.55, and 1727.58, the composition with the highest biological relevance was correlated with the higher mass accuracy calculation. For molecular ions at *m/z* 1271.47 and 1313.45, under the current instrumental conditions, both proposed compositions could be acceptable. For ions at *m/z* 1112.45 and 1208.47 both proposed compositions could not be accepted taking in account biosynthetic rules for the glycan assembly. This fact reflects a high necessity for future improvements of a biofilter library regarding the non-relevant glycan moieties. For more accurate compositional evaluation for these ions, high resolution MS and/or fragmentation analysis are necessary. For the first structural evaluation of MS data which can cover around 70% of compositions the SysBioWare platform offers a time scale of less than 1 min, in which the analysis of the raw mass spectrum along with the compositional assignment of glycoconjugate species can be accomplished.

APPLICATION TO GLYCOSPHINGOLIPIDOMICS

Glycosphingolipids (GSLs) are major components of the outer leaflet of the cell membrane, playing pivotal roles in a variety of biological processes, including the development of cancer [34, 35]. In course of distinct cell surface events during infection or tumour development they show disease-related expression changes. Thus, they could serve as useful targets for biomarker discovery. The molecular structure of GSLs is composed of a hydrophilic, highly variable carbohydrate chain and a lipophilic ceramide anchor. Both portions exhibit numerous structural variations, the combination of which results in a large diversity of GSL structures that can potentially exist [36]. Several hundred GSLs have already been characterized that differ only with regard to the glycan structures (http://sphingolab.biology.gatech.edu) and (http://www.glycosciences.de). The ceramide moiety exhibits variation in the number of carbon atoms, double bonds incorporated into the aliphatic chains, and hydroxyl groups present. Such variation of potential GSL structures, along with the biosynthetic rules, has to be taken into account upon data interpretation.

Applying SysBioWare to glycolipid analysis [37] the program was first adapted according to the structural specificities of GSLs, considering the glycan and the ceramide moiety as separate moieties, accordingly that the ceramide moiety can be introduced as a new building block. In humans, the ceramide is composed of a long chain base, the sphingosine, and a fatty acid bound to the amino group. The GSL model includes all possible structures with fatty acid substitutions from C 14 to C 26, where up to two double bonds can be assigned. The N-acyl linked fatty acids can be hydroxylated or non-hydroxylated and may have an odd number of C atoms. The so-called lyso-forms of GSLs, composed only of the glycan and the long chain base, i. e. without fatty acid substitution, are also supported by this model. The restrictive rules of the BioFilter were derived from the set of GSL structure found in humans (http://sphingolab.biology.gatech.edu).

MS data acquired from a complex GSL mixture purified from human serum were evaluated by the SysBioWare program, in which the Peak Detection Wizard was optimized for GSL analysis, achieving up to 94% of the true positive monoisotopic *m/z* values. By applying the BioFilter to the compositional analysis, it was possible to reduce the number of candidates by a factor of 2 to 4 for 75% of the cases, and by a factor of 5 to 9 for 18% of the cases. 14

peaks were unambiguously assigned to a single composition, whereas in 14 other cases, the correct compositions were placed at the highest rank when scored according to the mass accuracy.

Distinguishing between isomeric structures for correct assignment is a challenging task to be performed from MS data by the BioFilter toolbox. A fucose containing GSL with a hydroxy fatty acid substitution has the same elemental composition as the congruent GSL with a hexose instead of the fucose and a non-hydroxy fatty acid. These aspects were shown to be accurately solved by fragmentation analysis in MS/MS. The interpretation of MS/MS data was supported by *in silico* fragmentation module, calculating a comprehensive set of predicted fragment ions. This function was considered to be a powerful tool for the validation of proposed GSL structures.

According to the SysBioWare assignment proposal, several candidate structures for the signal at *m/z* 1249.74 would be possible, among them the isobaric structure variants with or without fucosylation and hydroxylated fatty acids. *In silico* fragmentation pattern of globotetraosylceramide (Gb4Cer) (d18:1, C16:0) $[M+Na]^+$ was in good agreement with that obtained by CID experiments of the precursor ions at *m/z* 1249.74, providing a high sequence coverage. Besides, the presence of low amounts of an isomeric or isobaric structure that is likely to contain a terminal hexose in contrast to Gb4Cer with a terminal *N*-acetylhexosamine unit was indicated by low abundant ions at *m/z* 1069.72 and *m/z* 1087.70, which can possibly be assigned to the isobaric neolactotetraosylceramide (nLc4Cer) (d18:1, C16:0) $[M+Na]^+$, a minor component of the human serum glycosphingolipidome [38]. Upon *in silico* fragmentation of the nLc4Cer component, 43 of the computed ions could be matched with the experimental data, many of which overlapping with fragment ions of its isobar, Gb4Cer.

Combined with *in silico* fragmentation matching, the potential of the SysBioWare platform was shown to be extendable also to GSL samples. Accordingly, it can be considered for rapid and accurate evaluation of complex data sets obtained from different patients' consortia. In protocols, where GSL profiles obtained from clinical samples by MS and MS/MS analysis SysBioWare may serve as a powerful tool for high-throughput glycolipidomics.

Conclusions and Outlook

Algorithms for monoisotopic *m/z* values recognition based the peak shape matching and isotopic grouping have been developed. Peak detection algorithm has been optimized for glycomics applications and successfully tested for mass spectra at different level of complexity. Tuned BioFilter toolbox, providing filtration of unreliable structures has been incorporated into Compositional Module. A novel software platform with an integrated Peak Detection Module and Compositions has been developed, providing the option for high-throughput data analysis within the time scale less than one minutes and their organization into the local data base for further analysis and mutual correlation. The developed software has been tested for the analysis of complex glycoconjugate mixtures from CDG patients and validated as a potential tool for clinical applications. Combined with *in silico* fragmentation matching, the potential of the SysBioWare platform was shown to be extendable to different glycoconjugate species.

The development of technologies for mass spectrometric analysis during the past 15 years opened new horizons for glycomics. Using modern mass spectrometry very small amounts of complex biological material can be analyzed in a very short time very accurately. Following the progress in instrumentation for data acquisition, huge amount of data are presently generated which require efficient ways to be classified and deposited in databases for applications in life sciences, biotechnology and medicine. Therefore is it high time to get new computational tools available for the broad community of scientists, which are not necessarily always trained as glycochemists or glycobiologists, but who will be using these glycomics tools to make significant contributions to our understanding of the function and the functionally or developmentally induced changes of carbohydrate structures in cells or subcellular compartments. To reach new frontiers in systems biology, glycomics will make progress by providing new and spectacular information, but the task in coming years will be to integrate the work on the fundamentals of this approach and to provide sufficient intellectual and financial resources for the integrated platforms.

Abbreviations

CDG	Congenital Disorders of Glycosylation
CID	Collision induced dissociation
Da	Dalton
dHex	Deoxyhexose
ESI	Electrospray ionization
Gal	Galactose
GlcNAc	*N*-acetylglucosamine
GSL	Glycosphingolipid
Hex	Hexose
HexNAc	*N*-acetylhexosamine
NeuGc	*N*-glycolylneuraminic acid
MALDI	Matrix-assisted laser desorption/ionization
Man5	Oligomannose

MS	Mass spectrometry
MS/MS (MS^n)	Tandem mass spectrometry (n stages)
m/z	Mass to charge ratio
NA2	Asialo-, galactosylated biantennary N-glycan
NeuAc	*N*-acetylneuraminic acid
NGA2	Asialo-, agalacto-, biantennary N-glycan
NGA2F	Asialo-, agalacto-, biantennary N-glycan with core fucose
ppm	Parts per million
Q	Quadrupole
r.i.	Relative intensity
S	Sulphate
Ser	Serine
Thr	Threonine
TOF	Time-of-flight

REFERENCES

[1] Egge, H., Peter-Katalinić (1987) *J. Mass Spectrom. Rev.* **6**:331 – 393.
doi: http://dx.doi.org/10.1002/mas.1280060302.

[2] Packer, N.H., von der Lieth, C.W., Aoki-Kinoshita, K.F., Lebrilla, C.B., Paulson, J.C., Raman, R., Rudd, P., Sasisekharan, R., Taniguchi, N., York, W.S. (2008) *Proteomics* **8**:8 – 20.
doi: http://dx.doi.org/10.1002/pmic.200700917.

[3] Aoki-Kinoshita, K.F. (2008) *PLoS Comput. Biol.* **4**.
doi: http://dx.doi.org/10.1371/journal.pcbi.1000075.

[4] von der Lieth, C.W., Bohne-Lang, A., Lohmann, K.K., Frank, M. (2004) *Brief Bioinform.* **5**:164 – 178.
doi: http://dx.doi.org/10.1093/bib/5.2.164.

[5] von der Lieth, C.W., Lutteke, T., Frank, M. (2006) *Biochim. Biophys. Acta* **1760**:568 – 577.
doi: http://dx.doi.org/10.1016/j.bbagen.2005.12.004.

[6] Cooper, C.A., Gasteiger, E., Packer, N.H. (2001) *Proteomics* **1**:340 – 349.
doi: http://dx.doi.org/10.1002/1615-9861(200102)1:2<340::AID-PROT340>3.3.CO;2-2.

[7] Cooper, C.A., Joshi, H.J., Harrison, M.J., Wilkins, M.R., Packer, N.H. (2003) *Nucleic Acids Res.* **31**:511 – 513.
doi: http://dx.doi.org/10.1093/nar/gkg099.

[8] Go, E.P., Rebecchi, K.R., Dalpathado, D.S., Bandu, M.L., Zhang, Y., Desaire, H. (2007) *Anal. Chem.* **79**:1708 – 1713.
doi: http://dx.doi.org/10.1021/ac061548c.

[9] Joshi, H.J., Harrison, M.J., Schulz, B.L., Cooper, C.A., Packer, N.H., Karlsson, N.G. (2004) *Proteomics* **4**:1650 – 1664.
doi: http://dx.doi.org/10.1002/pmic.200300784.

[10] Lohmann, K.K., von der Lieth, C.W. (2003) *Proteomics* **3**:2028 – 2035.
doi: http://dx.doi.org/10.1002/pmic.200300505.

[11] Lohmann, K.K., von der Lieth, C.W. (2003) *Glycobiology* **13**:846 – 846.

[12] Gaucher, S.P., Morrow, J., Leary, J.A. (2000) *Anal. Chem.* **72**:2331 – 2336.
doi: http://dx.doi.org/10.1021/ac000096f.

[13] Maass, K., Ranzinger, R., Geyer, H., von der Lieth, C.W., Geyer, R. (2007) *Proteomics* **7**:4435 – 4444.
doi: http://dx.doi.org/10.1002/pmic.200700253.

[14] Apte, A., Meitei, S.M. (2010) *Methods Mol. Biol.* **600**:269.
doi: http://dx.doi.org/10.1007/978-1-60761-454-8_19.

[15] Ethier, M., Saba, J.A., Ens, W., Standing, K.G., Perreault, H. (2002) *Rapid Commun. Mass. Spectrom.* **16**:1743 – 1754.
doi: http://dx.doi.org/10.1002/rcm.779.

[16] Ethier, M., Saba, J.A., Spearman, M., Krokhin, O., Butler, M., Ens, W., Standing, K.G., Perreault, H. (2003) *Rapid Commun. Mass Spectrom.* **17**:2713 – 2720.
doi: http://dx.doi.org/10.1002/rcm.1252.

[17] Goldberg, D., Sutton-Smith, M., Paulson, J., Dell, A. (2005) *Proteomics* **5**:865 – 875.
doi: http://dx.doi.org/10.1002/pmic.200401071.

[18] Goldberg, D., Bern, M., Parry, S., Sutton-Smith, M., Panico, M., Morris, H. R., Dell, A. (2007) *J. Prot. Res.* **6**:3995 – 4005.
doi: http://dx.doi.org/10.1021/pr070239f.

[19] Vakhrushev, S.Y., Dadimov, D., Peter-Katalinić, J. (2009) *Anal. Chem.* **81**:3252 – 3260.
doi: http://dx.doi.org/10.1021/ac802408f.

[20] Linden, H.U., Klein, R.A., Egge, H., Peter-Katalinić, J., Dabrowski, J., Schindler, D. (1989) *Biol. Chem. Hoppe Seyler* **370**:661 – 672.

[21] Schindler, D., Kanzaki, T., Desnick, R.J. (1990) *Clin. Chim. Acta* **190**:81 – 91. doi: http://dx.doi.org/10.1016/0009-8981(90)90282-W.

[22] Jaeken, J. (2004) *J. Inherit. Metab. Dis.* **27**:423 – 426. doi: http://dx.doi.org/10.1023/B:BOLI.0000031221.44647.9e.

[23] Jaeken, J., Carchon, H. (2004) *Curr. Opin. Pediatr.* **16**:434 – 439. doi: http://dx.doi.org/10.1097/01.mop.0000133636.56790.4a.

[24] Jaeken, J., Matthijs, G. (2007) *Annu. Rev. Genomics Hum. Genet.* **8**:261 – 278. doi: http://dx.doi.org/10.1146/annurev.genom.8.080706.092327.

[25] Frösch, M., Bindila, L., Zamfir, A., Peter-Katalinić, J. (2003) *Rapid Commun. Mass Spectrom.* **17**:2822 – 2832. doi: http://dx.doi.org/10.1002/rcm.1273.

[26] Vakhrushev, S.Y., Langridge, J., Campuzano, I., Hughes, C., Peter-Katalinić, J. (2008) *J. Clin. Proteom.* **4**:47 – 57. doi: http://dx.doi.org/10.1007/s12014-008-9010-3.

[27] Vakhrushev, S.Y., Langridge, J., Campuzano, I., Hughes, C., Peter-Katalinić, J. (2008) *Anal. Chem.* **80**:2506 – 2513. doi: http://dx.doi.org/10.1021/ac7023443.

[28] Vakhrushev, S.Y., Mormann, M., Peter-Katalinić, J. (2006) *Proteomics* **6**:983 – 992. doi: http://dx.doi.org/10.1002/pmic.200500051.

[29] Vakhrushev, S.Y., Snel, M.F., Langridge, J., Peter-Katalinić, J. (2008) *Carbohydrate. Res.* **343**:2172 – 2183. doi: http://dx.doi.org/10.1016/j.carres.2007.11.014.

[30] Vakhrushev, S.Y., Zamfir, A., Peter-Katalinić, J. (2004) *J. Am. Soc. Mass Spectrom.* **15**:1863 – 1868. doi: http://dx.doi.org/10.1016/j.jasms.2004.09.008.

[31] Zamfir, A., Vakhrushev, S., Sterling, A., Niebel, H.J., Allen, M., Peter-Katalinić, J. (2004) *Anal. Chem.* **76**:2046 – 2054. doi: http://dx.doi.org/10.1021/ac035320q.

[32] Vakhrushev, S.Y., Mormann, M., Peter-Katalinić, J. (2006) *Proteomics* **6**:983 – 992. doi: http://dx.doi.org/10.1002/pmic.200500051.

[33] Vakhrushev, S.Y., Zamfir, A., Peter-Katalinić, J. (2004) *J. Am. Soc. Mass Spectrom.* **15**:1863 – 1868.
doi: http://dx.doi.org/10.1016/j.jasms.2004.09.008.

[34] Hakomori, S. (1996) *Cancer Res.* **56**:5309 – 5318.

[35] Kannagi, R., Yin, J., Miyazaki, K., Izawa, M. (2008) *Biochim. Biophys. Acta* **1780**:525 – 531.
doi: http://dx.doi.org/10.1016/j.bbagen.2007.10.007.

[36] Peter-Katalinić, J., Egge, H. (1990) *Methods Enzymol.* **193**:713 – 733.
doi: http://dx.doi.org/10.1016/0076-6879(90)93446-R.

[37] Souady, J., Dadimov, D., Kirsch, S., Bindila, L., Peter-Katalinić, J., Vakhrushev, S.Y. (2010) *Rapid Commun. Mass Spectrom.* **24**:1 – 10.
doi: http://dx.doi.org/10.1002/rcm.4479.

[38] Kundu, S.K., Diego, I., Osovitz, S., Marcus, D.M. (1985) *Arch. Biochem. Biophys.* **238**:388 – 400.
doi: http://dx.doi.org/10.1016/0003-9861(85)90179-1.

Biographies

Ten Feizi

is director of the Glycosciences Laboratory at Imperial College London.

During 1970s–80s, working with the human monoclonal autoantibodies, and later with murine hybridoma antibodies, Feizi and her colleagues established that, among these were antibodies exquisitely specific for oligosaccharide sequences of the blood group family. Using these antibodies as sequence-specific reagents, she observed that there occur programmed changes in the expression and patterning of blood the group-related sequences during embryonic development and cell differentiation, and in oncogenesis. This led her to predict important roles, now corroborated, for oligosaccharides as recognition elements in molecular interactions.

With her colleagues, in 1985, Feizi developed the neoglycolipid (NGL) technology as a means of singling out and micro-sequencing ligand-bearing oligosaccharides recognized by carbohydrate-binding proteins. Coupled with mass spectrometry, the technology has proven to be uniquely powerful for carbohydrate ligand discovery, and since 2002, it is the basis of an advanced carbohydrate microarray platform with > 600 sequence-defined oligosaccharide probes.

The microarray system is continually expanding, aimed at thousands of probes, and is validated for generating 'designer' microarrays from targeted tissues and macromolecules and pinpointing ligands. Coupled with dedicated software, it is showing considerable promise as a novel approach to surveying entire glycomes and proteomes for the molecular definition of carbohydrate-recognition systems: endogenous, as well as those that play as part in pathogen-host interactions.

Pamela Greenwell

has worked for more than 30 years as a glycobiologist and as a molecular biologist for about 20 years. She completed her PhD at the MRC Clinical Research Centre in Harrow where she worked for 15 years, funded by the MRC and ICRF, under the guidance of Prof. Winifred M. Watkins and Prof. Walter T. Morgan. Pamela's PhD thesis investigated the biochemistry of glycosyltransferases that synthesized human blood groups. In 1989, on the closure of the department, she joined the University of Westminster to run the new MSc course in Medical Molecular Biology that celebrates its 20th birthday this year.

Pamela has supervised 14 PhD students to completion and currently supervise 3. She has published more than 70 papers and 80 conference proceedings and a book "Molecular Therapeutics: 21st century Medicine". She was a WestFocus Knowledge Exchange Fellow and her remit is to facilitate interaction with industry. In that context, Pamela has organized 3 industry-facing conferences on proteomics. In collaboration with colleagues, she also organizes a series of short courses in Molecular Biology.

Over the last year, Pamela has been leading the bioscience group involved in the JISC funded ProSim project that aims to make molecular modeling more accessible by designing web-based workflows and utilizing GRID computing to increase the speed of simulation enabling meaningful numbers of simulations to be run and analyzed. In this context, she ran an international workshop on modeling protein-glycan interactions funded by the BBSRC in April this year for 50 delegates (http://www.omii.ac.uk/news/news.jhtml?nid=206) and has been an invited speaker at 4 NGS (National Grid Computing Service) workshops.

Current Research/Ongoing Projects

- Data mining the trichomonad genome
- Development of tools for glycobiology, bioinformatics and molecular modelling
- Role of anaerobes and glycans in diabetic foot ulcers

Stuart Haslam

After obtaining my first degree and PhD from the University of Leeds, my postdoctoral research at Imperial College London on nematode parasite glycosylation laid the foundations of high sensitivity mass spectrometric analysis of complex mixtures of oligosaccharides which are now widely exploited in glycomic analysis. My more recent research has encompassed the structural analysis of glycoconjugates from diverse biological origins ranging from bacteria to humans. I was appointed to a lectureship in 2002 at Imperial College London and that year I was invited by the International Steering Committee of the Consortium for Functional Glycomics (CFG) to serve as Director of its Analytical Glycotechnology Core when this facility was re-located from San Diego to Imperial College as recommended by the CFG's International Advisory Board. The CFG is an NIH-funded large research initiative aiming to understand the role of carbohydrate-protein interactions in cell-cell communication.

Seven scientific cores (http://www.functionalglycomics.org/static/consortium/main.shtml) provide resources to the international scientific community. The Analytical Glycotechnology Core, which is the only core to be based outside the USA, is involved in glycomic profiling of mouse and human tissues and performing carbohydrate structure analysis on carbohydrate binding protein ligands and specific mouse and human cell lines. I am also a member of the

Steering Committee of the EuroCarbDB consortium which is funded by the European Commission 6th Framework Programme. This consortium aims to develop and distribute web-based carbohydrate data bases and I lead the Imperial College component which is responsible for MS activities. I was promoted to senior lecture at Imperial in 2008.

Hans Heindl

I am a trained MD and have been working for about 20 years in primary care in Austria. Since my medical biochemistry studies I have always been interested in questions of structural biology. In 2004 I took a sabbattical and started to study medical molecular biology at the University of Westminster in London under supervision of Dr. Pamela Greenwell, a well known glycobiologist. I earned my masters degree working on RNA interference in *T. vaginalis*. Since then I have collaborated with the CPC department of Westminster in establishing applications for biologists on their local grid computer and the NGS. In 2009 I took the Protein Crystallography course at Birkbeck College London UK.

My fields of interest are:

- Working with homology models (to date there are no crystal structures available) of *Trichomonas vaginalis* glucosidases especially family 47 enzymes (mannosidases) and sialidases. Those enzymes play a crucial role in the maturation of glycosidic side chains of proteins. N-glycosylation is important in molecular recognition and modulates enzymatic activity. Glycosylation of a nascent protein chain may also help in folding and detection of misfolding. Evaluation of model quality. Sialidases seem to be crucial in the digestion of mucus and may play an important role in the pathogenesis of *T. vaginalis.*
- Using bioinformatics to find the candidate sequences in *T. vaginalis* (the genome of the organism has been completely sequenced, but many of the open reading frames have not yt been properly annotated.
- Trying to fit (dock) candidate substrates (*e.g.* sugars, olicosaccarides or putative inhibitors) to the macromolecular models of the enzymes using the *in-silico* docking algorithms of Autodock4.
- Using a molecular dynamics simulation (Amber 10.0 or Gromacs 4) to analyse the interactions between enzyme and substrate over the time, varying pressure and temperature.
- Using the insights gained for the planning of wet lab work (*e.g.* aminoacid substitutions).
- The simulations are greatly facilitated by the use of the NGS ressources.

Martin Hicks

is a member of the board of management of the Beilstein-Institut. He received an honours degree in chemistry from Keele University in 1979. There, he also obtained his PhD in 1983 studying synthetic and theoretical approaches to the photochemistry of pyridotropones under the supervision of Gurnos Jones. He then went to the University of Wuppertal as a post-doctoral fellow, where he carried out research with Walter Thiel on semi-empirical quantum chemical methods. In 1985, Martin joined the computer department of the Beilstein-Institut where he worked on the Beilstein Database project. His subsequent activities involved the development of cheminformatics tools and products in the areas of substructure searching and reaction databases.

Thereafter, he took on various roles for the Beilstein-Institut, including managing directorships of subsidiary companies and was head of the funding department 2000 – 2007. He joined the board of management in 2002 and his current interests and responsibilities range from organization of Beilstein Symposia with the aim of furthering interdisciplinary communication between chemistry and neighbouring scientific areas, to the publishing of Beilstein Open Access journals such as the Beilstein Journal of Organic Chemistry and the Beilstein Journal of Nanotechnology.

Carsten Kettner

studied biology at the University of Bonn and obtained his diploma at the University of Göttingen. Here, in Prof. Gradmann's group "Molecular Electrobiology" which consisted of people carrying out research in electrophysiology and molecular biology in fruitful cooperation, he studied transport characteristics of the yeast plasma membrane using patch clamp techniques. In 1996 he joined the group of Dr. Adam Bertl at the University of Karlsruhe and successfully narrowed the gap between the biochemical and genetic properties, and the biophysical comprehension of the vacuolar proton-translocating ATP-hydrolase. He was awarded his Ph.D for this work in 1999. As a post-doctoral student he continued both the studies on the biophysical properties of the pump and investigated the kinetics and regulation of the dominant plasma membrane potassium channel (TOK1). In 2000 he moved to the Beilstein-Institut to represent the biological section of the funding department. Here, he is responsible for the organization of the Beilstein symposia, research (proposals) and publication of the proceedings of the symposia. Since 2004 he coordinates the work of the STRENDA commission and promotes along with the commissioners the proposed standards of reporting enzyme data. In 2007 he became involved in the development of a program for the establishment of Beilstein Endowed Chairs for Chemical Sciences and related sciences. At the same time he started a correspondence course at the Studiengemeinschaft Darmstadt (a certified service provider) where he was awarded his certificate of competence as project manager for his studies and thesis. In 2009, he took part in the establishment of a large-scale funding project in the nanoscience area called NanoBiC.

Oliver Kohlbacher

studied chemistry and computer science at Saarland University, Saarbrücken. After a Diplom in chemistry, he obtained a PhD in computer science (Max Planck Institute for Informatics, Saarbrücken). He headed an independent research group on protein docking at Saarland University between 2000 and 2003. He spent time as a postdoc in the group of Eugene Myers at Celera Genomics in 2001 and 2002, where he was involved with the Celera Proteomics Factory.

In 2003, he took up the position of full professor for Simulation of Biological Systems at the Center for Bioinformatics at Eberhard Karls University Tübingen. His research interests are focused on the application of computational methods in proteomics, immunomics, and drug design. In drug design, his key interests are currently the development of accurate empirical scoring functions and hybrid docking-QSAR approaches.

Hiroshi Mamitsuka

received his B.S. in biophysics and biochemistry, M.E. in information engineering and Ph.D in information sciences, all from University of Tokyo, Japan. He worked with NEC Research Laboratories in Japan for eleven years, being engaged in research on machine learning with a variety of applications including bioinformatics and business data mining. He then joined Institute for Chemical Research (ICR) of Kyoto University, where he is currently a professor being jointly appointed as a faculty of the Graduate School of Pharmaceutical Sciences of the same university. His current research interests are, broadly speaking, molecular biology and chemical genomics, based on machine learning and data mining approaches.

Hannu Peltoniemi

has background on physics and mathematics that he studied at Helsinki University of Technology, Finland, and graduated (M.Sc.) at 1996. Until 2002 he worked on algorithm and software development of environmental fluid flow and wave problems at Environmental Impact Assessment Centre of Finland Ltd and VTT Technical Research Centre of Finland. The application area shifted to biology when he moved to Medicel Ltd, Finland, to take part of the systems biology platform development. The main area was tandem mass spectrometry data-analysis, especially development of novel tools for *de-novo* glycopeptide analysis.

Since 2008 he has been an independent consultant through his company Applied Numerics Ltd, with special emphasis on glycan analysis.

Jasna Peter-Katalinić

is professor emeritus of biophysics at the University of Münster, Germany. Born and educated in Zagreb, Croatia, she obtained PhD in organic chemistry at the University of Zürich, Switzerland. After the postdoc time at the Texas A&M University, USA, she was employed as a scientist at the University of Bonn, Germany, where she obtained the habilitation in physiological chemistry. She pioneered the introduction of modern mass spectrometric methods to structural glycobiology/glycomics, as described in more than 280 publications in journals and books. After her move to University of Münster, she was the founder of the laboratory "Biomedical Analysis" at the Medical School and its director 1996 – 2008.

She was the board member of the German Society for Mass Spectrometry 2000 – 2008 and a member of the Human Glycoproteome Initiative (HGPI) from HUPO 2004 – 2008.

In 2002 she received the 1st Life Science Award from the German Society of Mass Spectrometry (DGMS). She was visiting professor at the Medical College of Georgia, USA, and at the Eidgenössische Technische Hochschule, Switzerland. She is currently an associated editor of *Journal of Proteomics* and a member of the Advisory Board for the *Journal of the American Society of Mass Spectrometry* and a guest researcher in the Institute for Pharmaceutical Biology and Phytochemistry at the University of Münster. Since 2001, she is the chair organizer of the ongoing series of "Summer Courses on Mass Spectrometry in Biotechnology and Medicine in the Center for Advanced Academic Studies, Dubrovnik, Croatia".

Her current interests are in developing integrated instrumental platforms for human glycoproteomics and nanobioanalytics for complex carbohydrates, including new software for automatic assigment of mass spectrometric data.

Peter H. Seeberger

received his Vordiplom in 1989 from the Universität Erlangen-Nürnberg, where he studied chemistry as a Bavarian government fellow. In 1990 he moved as a Fulbright scholar to the University of Colorado where he earned his Ph.D. in biochemistry under the guidance of Marvin H. Caruthers in 1995. After a postdoctoral fellowship with Samuel J. Danishefsky at the Sloan-Kettering Institute for Cancer Research in New York City he became Assistant Professor at the Massachusetts Institute of Technology in January 1998 and was promoted to Firmenich Associate Professor of Chemistry with tenure in 2002. From June 2003 until January 2009 he held the position of Professor for Organic Chemistry at the Swiss Federal Institute of Technology (ETH) in Zürich, Switzerland where he served as chair of the laboratory in 2008.

In 2009 he assumed positions as Director at the Max-Planck Institute for Colloids and Surfaces in Potsdam and Professor at the Free University of Berlin. Since 2003 he serves as an Affiliate Professor at the Burnham Institute in La Jolla, CA.

Professor Seeberger's research has been documented in over 215 articles in peer-reviewed journals, two books, fifteen issued patents and patent applications, more than 100 published abstracts and more than 450 invited lectures. Among other awards he received the Arthur C. Cope Young Scholar and Horace B. Isbell Awards from the American Chemical Society (2003), the Otto-Klung Weberbank Prize for Chemistry (2004), the Havinga Medal (2007), the Yoshimasa Hirata Gold Medal (2007), the Körber Prize for European Sciences (2007), the UCB-Ehrlich Award for Excellence in Medicinal Chemistry (2008) and the Claude S. Hudson Award for Carbohydrate Chemistry from the American Chemical Society (2009). In 2007 and 2008 he was selected among "The 100 Most Important Swiss" by the magazine "Schweizer Illustrierte". In 2010 he will receive the Tetrahedron Young Investigator Award for Bioorganic and Medicinal Chemistry.

Peter H. Seeberger is the Editor of the *Journal of Carbohydrate Chemistry* and serves on the editorial advisory boards of seven other journals. He is a founding member of the board of the *Tesfa-Ilg "Hope for Africa" Foundation* that aims at improving health care in Ethiopia in particular by providing access to malaria vaccines and HIV treatments. He is a consultant and serves on the scientific advisory board of several companies. In 2006 he served as president of the Swiss Academy of Natural Sciences.

The research in professor Seeberger's laboratory has resulted in two spin-off companies: Ancora Pharmaceuticals (founded in 2002, Medford, USA) that is currently developing a promising malaria vaccine candidate in late preclinical trials as well as several other therapeutics based on carbohydrates and i2chem (2007, Watertown, USA) that is commercializing microreactors for chemical applications.

Mark S. Stoll

I am a graduate in Organic Chemistry and have been a research worker for my entire career. I have worked in the Glycosciences Laboratory for the past 25 years and have participated in the development of the 'Neoglycolipid Technology', developed over the last 20 years. This technology has greatly facilitated the study of the interactions between carbohydrate-binding proteins and their target glycoconjugates. The development has passed through a number of steps to become the powerful tool that it is today.

I am currently involved in further extending the technology to a new level of miniaturisation using 'Microarray Technology'. Briefly, this consists of 'printing' femtomol quantities of neoglycolipid probes on specially prepared glass slides as a large array of spots each of which can be targeted by antibodies and other carbohydrate binding proteins and then their binding measured by laser fluorometry. This technology produces very large amounts of data that must be processed to yield accessible results. I have developed computer software to transform raw fluorescence data, coupled with carbohydrate information stored in our local database, which I designed, into tables, charts and other graphical representations to facilitate understanding of the binding experiments. My hope now is to establish collaborative links with bioinformatics experts to render the software Web-based and accessible to other groups carrying out carbohydrate microarray analyses.

Index of Authors

Index